ADVANCES IN POLYMER SCIENCE AND ENGINEERING

ADVANCES IN POLYMER SCIENCE AND ENGINEERING

Proceedings of the Symposium on Polymer Science and Engineering held at Rutgers University, October 26-27, 1972

Edited by

K. D. Pae, D. R. Morrow, and Yu Chen
Department of Mechanics and Materials Science
College of Engineering
Rutgers University
The State University of New Jersey
New Brunswick, New Jersey

PLENUM PRESS • NEW YORK - LONDON • 1972

Library of Congress Catalog Card Number 72-88423

ISBN 0-306-30713-8

© 1972 Plenum Press, New York
A Division of Plenum Publishing Corporation
227 West 17th Street, New York, N. Y. 10011

United Kingdom edition published by Plenum Press, London
A Division of Plenum Publishing Company, Ltd.
Davis House (4th Floor), 8 Scrubs Lane, Harlesden, London,
NW10 6SE, England

Printed in the United States of America

PREFACE

The Proceedings and the Symposium on Polymer Science and Engineering, to be held on October 26 and 27, 1972 at Rutgers University, are in honor of Professor John A. Sauer. October 26, 1972 marks the 60th birthday of Professor Sauer and we feel it is quite appropriate to make note of this event. All of the contributing authors have eagerly submitted their original works as an expression of their esteem and affection for this dedicated man, friend, husband, father, scientist and teacher.

This book could have been made extremely voluminous and the Symposium could have gone on for days. However, the achievements of a man such as Jack Sauer do not have to be measured by the number of pages in a book nor the number of speakers at a meeting. A more meaningful measure is the sincerity and devotion with which these few pages were assembled. All of the contributions to these Proceedings are from invited speakers. Numbered among the contributors are some of Jack's many personal friends as well as numerous former students who are currently working in the field of polymer science and engineering. It will be apparent to all who know him that those included represent but a small portion of Jack's friends and students. Although a fairly exhaustive search was made before the invitations were sent the number of omissions is both extremely large and unavoidable. To those many friends and former students whom we have missed, our sincerest apologies are hereby offered.

The title of the Symposium and of these Proceedings has been selected to reflect the current nature of each individual contribution. No effort has been made to unify the writing style, or the format by which these papers have been presented. In this way we have attempted to preserve the personal nature of each presentation.

On behalf of all of his friends, we wish to say "Happy Birthday, Jack and Thanks!"

Y. Chen
D. R. Morrow
K. D. Pae

June 30, 1972

DEDICATION

The dedication of these Proceedings and the Symposium on Polymer Science and Engineering to Professor John A. Sauer in honor of his 60th birthday is a fitting tribute by his friends and former students. It is not really necessary to belabor his contributions in the field of polymer science and engineering, and his great contribution to the advancement of science and humanity. Jack's devotion to the education of young men and women and his deep interest and understanding of people are well-recognized and appreciated. In dedicating this volume as a small token of our esteem and appreciation we wish him continued success in all his endeavors and a happy 60th birthday.

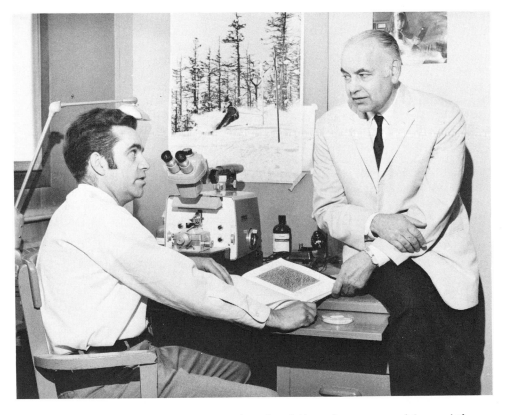

Professor John A. Sauer doing what he likes best - working with a student.

BIOGRAPHY OF JOHN A. SAUER

Jack Sauer was born in Elizabeth, New Jersey sixty years ago. He received his B.S. and M.S. degrees from Rutgers University in 1934 and 1936, respectively, and was awarded a Ph.D. degree in Mathematical Physics by Cambridge University in 1942.

Jack has held a variety of positions in institutes, in industry and at universities. He was an Instructor in Mechanics at Carnegie Institute of Technology in 1940-41, an Industrial and Senior Fellow at Mellon Institute in 1941-44 and an Assistant to the President and then Director of Research and Engineering at the Elastic Stop Nut Corporation in 1944-45. From there Jack took the position as Chairman of the Engineering Mechanics Department at The Pennsylvania State University. In 1952-53 he was a Visiting Professor at the Clarendon Laboratory at Oxford. Upon his return to Penn State he assumed the chairmanship of the Physics Department. He was a Guggenheim Fellow in 1959-60, spending the year in Oxford. In 1963 he left Penn State and returned to Rutgers as Chairman of the Mechanics and Materials Science Department.

Although for many years Jack has been involved in administration, he has always been an active teacher and researcher. He has taught graduate and undergraduate courses in polymer science, engineering and physics. He is an author or coauthor of over 100 research papers on fracture, nuclear magnetic resonance, dynamic mechanical behavior and single crystal morphology of high polymers; he has also authored numerous monographs on these subjects. His contributions continue to lead the way in the fascinating and complex field of polymer science and engineering.

SELECTED LIST OF PUBLICATIONS
by John A. Sauer

Book

    Strength of Materials, 2nd Edition
    (Co-author J. Marin), McMillan Company, 1954

Selected Papers

    Magnetic Energy Constants of Dipolar Lattices, Phys. Rev. 57,
142 (1940).

    Theoretical Study of a Possible Model of Paramagnetic Alums at
Low Temperatures (co-author, H.N.V. Temperley), Proc. Roy. Soc.
A 176, 203 (1940).

    On the Adiabatic Demagnetization of Iron Alum, Phys. Rev. 64,
94 (1943).

    Theory of the Elastic Stop Nut, booklet published by ESNA,
(1943).

    Relative Temperature Stability of Stressed Plastics (co-
author F. A. Schwertz, D. L. Worf), Modern Plastics 22, Mar. 1945.

    Fatigue of Aluminum Under Combined Stress, Proc. 7th Int. Cong.
for Appl. Mech. 5, 150 (1948).

    Obtaining Fatigue Test Data (co-author P. K. Roos), Mach. Des.
20, 115, 158 (1948).

    Creep and Damping Properties of Polystyrene (co-authors J.
Marin, C. C. Hsiao), J. App. Phys. 20, 507 (1949).

    Effect of Steady Stress on Fatigue Behavior of Aluminum (co-
author D. C. Lemmon), Trans. Amer. Cos. Metals 42, 559 (1949).

    Damping and Resonant Load-Carrying Capacities of Polystyrene
and Other High Polymers (co-author, J. Oliphant), Proc. Amer. Soc.
Test. Mat. 49, 1119 (1949).

    Vibration, Colliers Encyclopedia (1949).

    New Type of Course in Engineering materials (co-author J. Marin),
J. Eng. Ed. 41, 81 (1950).

    On Crazing of Linear High Polymers (co-author C. C. Hsiao),
J. Appl. Phys. 21, 1071 (1950).

Bolts – How to Prevent Their Loosening (co-authors D. C. Lemmon, E. K. Lynn), Mach. Des. <u>22</u>, 133 (1950).

Effect of Strain Rate on the Tensile and Compressive Stress-Strain Properties of Polystyrene (co-author C. C. Hsiao), Amer. Soc. Testing Mat. Bull. No. 172, p. 29 (1951).

Research in Mechanics and Materials (co-author J. Marin), Eng. Res. Revs. <u>1</u>, 12 (1952).

Stress Crazing of Plastics (co-author C. C. Hsiao), India Rubber World 128, 355 (1953).

Effect of Biaxial Orientation on the Ultimate Strength and Crazing of Linear Polymers (co-author C. C. Hsiao), J. App. Phys. 24, 957 (1953).

Plastic Stress-Strain Relations under Radial and Non-Radial Combined Stress Loading (co-author J. Marin), J. Franklin Inst. <u>256,</u> 119 (1953).

Determination of Internal Friction in Metals (co-author J. N. Brennan), Proc. 8th Int. Cong. on Theo. and Appl. Mech. 426 (1953).

Stress-Crazing of Plastics (co-author C. C. Hsiao), Trans. Amer. Soc. Mech. Eng. P. 895 (1953).

Stress-Crazing of Plastics (co-author C. C. Hsiao), Soc. Plast. Eng. J. <u>9</u>, (1953).

Behavior of Aluminum under Biaxial Stress (co-author J. Marin), Proc. 8th Int. Congress on Theoretical and Applied Mechanics 213 (1953).

Energy Dissipation in Vibrating Solids, Virginia J. of Sci. 5 (1954).

Dispersion of Ultrasonic Pulse Velocity in Cylindrical Rods (co-authors J. N. Brennan, L.Y. Tu), J. Accous. Soc. Am. <u>27</u>, 550 (1955).

Physics for Engineers (co-author D. E. Hardenbergh), J. Eng. Ed. 46, 349 (1955).

Dynamical Mechanical Properties of Polystyrene, Polyethylene and Polytetrafluorethylene at Low Temperatures (co-author D. E. Kline), J. Poly. Sci., <u>18</u>, 491 (1955).

Relaxation Properties of High Polymers (co-author D. E. Kline), Ninth International Congress on Theoretical and Applied Mechanics, Brussels, September 1956, Proc. 5, 368 (1957).

Effect of Branching on Dynamic Mechanical Properties of Polyethylene (co-authors D. E. Kline and A. E. Woodward), J. Poly. Sci., 22, 455 (1956).

Proton Resonance in Neutron and Gamma-Irradiated Polyethylene (co-author N. Fuschillo), J. of Chem. Phys. 26, 1348 (1957).

Nuclear Magnetic Resonance and Crystallinity in Polyethylene (co-authors Elliot Rhian, N. Fuschillo), J. Polymer Sci. 25, 381 (1957).

Effect of Irradiation on Dynamic Mechanical Properties of 6-6 Nylon (co-authors C. W. Deeley, A. E. Woodward), J. Appl. Phys. 28, 1124 (1957).

Effect of Thermal History on the Dynamic Modulus at 20°C of Irradiated Polyethylene (co-authors A. E. Woodward, C. W. Deeley, D. E. Kline), J. Polymer Sci. 26, 383 (1957).

Effect of Pile Irradiation on the Dynamic Mechanical Properties of Polyethylene (co-authors C. W. Deeley, D. E. Kline, A. E. Woodward), J. Poly. Sci. 28, 109 (1958).

The Dynamic Mechanical Properties of High Polymers at Low Temperatures (co-author A. E. Woodward), Advances in Polymer Science 1, 114 (1958).

Dynamic Mechanical Behavior of Irradiated Polyethylene (co-authors C. W. Deeley, A. E. Woodward), J. Appl. Phys. 29, 1415 (1958).

Segmental Motion in Polypropylene (co-authors R. A. Wall, N. Fuschillo, A. E. Woodward), J. Appl. Phys. 29, 1385 (1958).

Segmental Motion in Polymers Below 200°K (co-authors N. Fuschillo, C. W. Deeley and A. E. Woodward), Proc. Fifth Int. Conf. on Low Temperature Physics and Chemistry Univ. of Wisconsin Press p. 608 (1958).

The Dynamic Mechanical Behavior of Polystyrene: Atactic and Isotactic (co-authors R. A. Wall, A. E. Woodward), J. Polymer Sci. 35, 281 (1959).

Segmental Motion in Polybutene-1 (co-authors A. E. Woodward, R. A. Wall), J. Chem. Phys. 30, 854 (1959).

The Dynamic Mechanical Behavior of Partially Crystalline Polymers (with A. E. Woodward), Soc. Chem. Ind. Monograph No. 5 (The Physical Properties of Polymers), p. 245 (1959).

Nuclear Magnetic Resonance and Thermal Expansion in Isotactic Polypropylene, Polybutene and Polypentene(co-authors A. E. Woodward, N. Fuschillo), J. App. Phys. 30, 1488 (1959).

Proton Magnetic Resonance of Poly(Hexamethylene Adipamide) (co-authors R. E. Glick, R. P. Gupta, A. E. Woodward), J. Polymer Sci. 42, 271 (1960).

Transitions in Polymers by Nuclear Magnetic Resonance and Dynamic Mechanical Methods (with A. E. Woodward), Rev. Mod. Phys. 32, 88 (1960).

Proton Magnetic Resonance of Some Polyamides (co-authors A. E. Woodward, R. E. Glick, R. P. Gupta), J. Polymer Sci. 45, 367 (1960).

Investigation of The Dynamic Mechanical Properties of some Polyamides (with A. E. Woodward and J. M. Crissman) J. Polymer Sci. 44, 23, (1960)

Proton Magnetic Resonance of Irradiated Polyethylene (co-authors R. E. Glick, R. P. Gupta, A. E. Woodward), Polymer 1, 340 (1960).

Dynamic Mechanical Studies of Irradiated Polyethylene (co-authors L. J. Merrill, A. E. Woodward), Polymer 1, 35 (1960).

Dynamic Mechanical Behavior of Some Partially Crystalline Poly -α-Olefins (co-authors A. E. Woodward, R. A. Wall), J. Polymer Sci. 50, 117 (1961).

Proton Magnetic Resonance of Some Poly-α-Olefins and Alpha Olefin Monomers (co-authors A. E. Woodward, A. Odajima), J. Phys. Chem. 65, 1384 (1961).

Proton Magnetic Resonance of Some α-Methyl Group-Containing Polymers and Their Monomers (co-authors A. Odajima and A. E. Woodward), J. Polymer Sci. 55, 181 (1961).

Dielectric Loss in Poly(Hexamethylene Adipamide) and Poly (hexamethylene Sebacamide) at Low Temperatures (co-authors N.M. Stein, R. G. Lauttman, A. E. Woodward), J. App. Phys. 32, 2352 (1961).

Effect of Radiation and Moisutre on the Dynamic Mechanical Properties of Polyethylene Terephthalate (co-author D. E. Kline), Polymer 2, 401 (1961).

Radiation and Moisture Effects in Polymerized Epoxy Resins (co-author D. E. Kline), Soc. Plastic Engs. Trans. 2, 21 (1962).

Dynamic Elastic Behavior of High Polymers at Audio Frequencies Soc. Plastic Engs. Trans. 2, 57 (1962).

Molecular Motion in Polystyrene and Some Substituted Polystyrenes (co-authors A. Odajima, A. E. Woodward), J. Polymer Sci. 57, 107 (1962).

Proton Magnetic Resonance of Some Synthetic Polypeptides (co-authors J. A. E. Kail, A. E. Woodward), J. Phys. Chem. 66, 1292 (1962).

Proton Magnetic Resonance of Some Normal Paraffins and Polyethylene (co-authors A. Odajima, A. E. Woodward) J. Phys. Chem 66, 718, 1962.

Dynamic Mechanical Studies of Irradiated Polypropylene (co-authors L. J. Merrill, A. E. Woodward), J. Polymer Sci. 58, 19 (1962).

Nuclear Radiation Effects in Polytetrafluoroethylene (co-author D. E. Kline), J. Polymer Sci. Part A, 1, 1621 (1963).

Effects of High Energy Irradiation on Polypropylene (co-authors S. E. Chappell, A. E. Woodward), J. Polymer Sci. Part A 1, 2805-2818 (1963).

Low Temperature Internal Friction in Nylon-4 (co-authors K. D. Lawson, A. E. Woodward), J. App. Phys. 34, 2492 (1963).

Dynamic Shear Behavior of Poly(γ-Benzyl-L-Glutamate), Poly (DL-Propylene Oxide) and Poly(ethyl vinyl either) co-authors R. G. Saba, A. E. Woodward), J. Polymer Sci. Part A 1, 1483-1490 (1963).

Proton Magnetic Resonance of Polyethylene Crystals, (co-authors J.A.E. Kail, A. E. Woodward), Polymer 4, 413 (1963).

Dynamic Mechanical Behavior of Some High Polymers from 6°K (co-authors J. M. Crissman, A. E. Woodward), J. Polymer Sci. Part A, 2, 5075 (1964).

Effect of Nuclear Radiation on the Thermal Conductivity of Polyethylene (co-authors J. N. Tomlinson, D. E. Kline), SPE Transactions, January, p. 44 (1965).

Mechanical Relaxation Phenomena (co-author A. E. Woodward), Chapter 7 of Physics and Chemistry of the Organic Solid State, pp. 637-723, Interscience (1965).

Mechanical Relaxation Phenomena (co-author A. E. Woodward), Addendum to Vol. II, Chapter 7 of Physics and Chemistry of the Organic Solid State, pp. 925-947, Interscience (1965).

Dilute Solution-Grown Polypropylene Single Crystals (co-authors D. R. Morrow, A. E. Woodward), Polymer Letters 3, 463 (1965).

Relaxation Phenomena in High Polymers (co-author A. E. Woodward), J. Polymer Sci. C 8, 137-142 (1965) Part C, No. 8

Dynamic Mechanical Behavior of Polystyrene and Related Polymers at Temperatures from 4.2°K (co-authors J. M. Crissman, A. E. Woodward), J. Polymer Sci. A 3, 2693 (1965).

Morphology of Solution-Grown Polypropylene Crystal Aggregates (co-authors D. R. Morrow, G. C. Richardson), J. Appl. Phys. 36, 3017 (1965).

Proton Magnetic Resonance of Poly(ethyl Acrylate), Poly(butyl Acrylate), Poly(ethyl Vinyl Ether), Poly(Vinyl Propionate), Poly(vinyl Stearate) and Vinyl Stearate from 77°K (co-authors L. J. Merrill, A. E. Woodward), J. Polymer Sci., A 3, 4243 (1965)

Morphology of Bulk Polypropylene (co-authors D. R. Morrow and K. D. Pae), Nature 211, 514, 1966.

Fracture Studies of Polypropylene (co-authors K. D. Pae and D. R. Morrow), Proc. 1st International Conference on Fracture Sendai, Japan, September 1965, 2, 1229, 1966.

Morphology and Fracture in Polypropylene (co-authors D. R. Morrow and K. D. Pae) Preprints, IUPAC International Symposium on Macromolecular Chemistry, Tokyo, September 1966.

Effect of Nuclear Radiation on the Mechanical Relaxation Behavior of Polytetrafluoroethylene (co-authors G. A. Bernier and D. E. Kline) J. Macromol. Sci. (Physics) B1, 335 (1967).

Mechanical Relaxation in Polymers, in Mechanics of the Solid State, University of Toronto Press (1968).

Effects of Thermal History on Isotactic Polypropylene (co-author K. D. Pae), J. Appl. Polymer Sci. 12, 1901-1919 (1968).

Effects of Thermal History on Polypropylene Fractions (co-author K. D. Pae), J. Appl. Polymer Sci. 12, 1921-1938 (1968).

Morphology of Solution-Crystallized Polypropylene (co-authors D. R. Morrow and A. E. Woodward), J. Polymer Sci., Part C, No. 16, 3401-3411 (1968).

Stress-Strain Behavior of Polypropylene under High Pressure (co-authors K. D. Pae and D. R. Mears), Polymer Letters 6, 773-778 (1968).

Structure and Thermal Behavior of Pressure-Crystallized Polypropylene (co-author K. D. Pae), J. Appl. Phys. 39, 4959-4968 (1968).

Morphology and Structure of Polymer Single Crystals Annals, N.Y. Acad. Sci. 155, 517-538 (1969).

Deformation of Polypropylene Single Crystals (co-authors P. Cerra and D. R. Morrow) J. Macromol. Sci. - Physics B3 (1), 33-51 (1969).

Relaxation Behavior of Polymers at Low Temperature (co-author R. Saba) J. Macromol. Sci. - Chem. A3(7), 1217-1255, (1969) (Nov.)

Effects of Hydrostatic Pressure on the Mechanical Behavior of Polyethylene and Polypropylene (co-authors D. R. Mears and K. D. Pae) J. Appl. Phys. 40, 4229-4237, (1969).

Plastic Deformation and Fracture of Composite Polypropylene Crystals (co-authors D. R. Morrow and R. Jackson) ACS Polymer Preprints 10(2) 929, (1969).

Effects of Hydrostatic Pressure on the Mechanical Behavior of Polytetrafluoroethylene and Polycarbonate (co-authors D. R. Mears and K. D. Pae) European Polymer Journal 6, 1015-1032 (1970).

Low Temperature Mechanical Relaxation in Poly (-p-xylylenes) (co-author C. Chung) Polymer 11, 454-461 (1970).

Stress–strain Temperature Relations in High Polymers (co-author A. E. Woodward) Chap. 3, "Polymer Thermal Analysis" p. 107–224 M. Dekker (1970).

Low Temperature Mechanical Relaxation in Polymers containing Aromatic Groups (co-author C. Chung) J. Poly. Sci. May(1970).

The Influence of Structure and Other Factors on Molecular Motions in Solid Polymers from 4°K to 300°K, Polymer Science Symposium No. 32, 69–122, J. Wiley (1971).

Mechanical Relaxation in Radiation–Polymerized Vinyl Stearate (co-author T. Lim); Polymer Preprints, ACS, $\underline{12}$, 547 (1971).

Mechanical Relaxation in Radiation–Polymerized Vinyl Stearate (co-author T. Lim); J. Macromol. Sci. Phys. $\underline{36}$(1), 191(1972).

The Influence of Molecular Weight on the Fatigue Behavior of Polystyrene (co-authors E. Foden and D. R. Morrow) J. Appl. Polymer Sci., $\underline{16}$, 527 (1972).

Influence of Pressure on the Elastic Modulus, the Yield Strength, and the Deformation of Polymers, (co-author K. D. Pae) (accepted for publication in Book "Solids under Pressure" Inst. of Mech. Engs. London).

CONTENTS

PLASTIC DEFORMATION OF UNORIENTED CRYSTALLINE POLYMERS UNDER TENSILE LOAD

A. Peterlin

Camille Dreyfus Laboratory, Research Triangle Institute

P. O. Box 12194, Research Triangle Park, N. C.   27709

Summary

In the tensile loading of the unoriented crystalline polymer the sample after a very small elastic range deforms plastically without a significant change of morphology up to the point where the stacked crystalline lamellae start fracturing and transforming into bundles of microfibrils of the new highly oriented fibrous structure. The thin micronecking zones where this transformation occurs are more or less randomly scattered in the neck (cold drawing) or all over the sample (hot drawing). During this process the long period, i.e., the length of crystal blocks with the intervening amorphous layers of the microfibril, becomes adjusted to the temperature of drawing increased by adiabatic local heating. The heating is caused by the work of plastic deformation during the destruction of lamellae. The effect is conspicuous only in the case where the long period of the unoriented polymer is larger than the long period of the drawn sample. Such an adjustment is not a consequence of melting and recrystallization or annealing because there is neither enough energy for melting nor annealing ever reduces the long period. The new fibrous material has a significantly larger elastic modulus and stress to draw than the original spherulitic material. This strain hardening is caused by the new microfibrillar morphology. As a consequence of the great many almost completely extended tie molecules connecting consecutive crystal blocks and originating from chain unfolding during lamella fracturing the microfibril has a very high longitudinal strength. The plastic deformation of the drawn polymer hence occurs without yielding by sliding motion of microfibrils which is opposed by van der Waals forces in the boundary between them. Although these forces are weaker than between adjacent

1

chains in the crystal lattice the large surface to volume ratio of
microfibrils with a diameter between 100 and 200Å and a length of
some ten microns yields a high stress to draw and stress to break.

Introduction

The study of deformation of single volume elements suggests
three stages of plastic deformation: (1) plastic deformation of
spherulitic structure, (2) transformation of spherulitic into
fibrous structure by fracture of lamellae and formation of microfibrils, (3) plastic deformation of fibrous structure. The first
stage taking place before the neck is very nearly the same for all
volume elements and proceeds so far that the lamellae by sliding
motion, rotation, chain tilt and slip are ready for fracture.
The second stage is the most drastic, changing completely the
morphology. The transformation occurs in narrow destruction zones
scattered more or less randomly over the neck in cold drawing and
over the whole sample in hot drawing. The third stage does not
change the fibrous morphology. Either the individual microfibrils
or the bundles of microfibrils may slide past each other up to the
point where the rapidly growing or coalescing point vacancy defects
of the microfibrillar superlattice, rupture of microfibrils and
longitudinal voids create sufficiently large critical cracks
leading to sample failure. The highly drawn sample usually breaks
in the characteristic fibrous manner with individually broken
fibrils protruding the fracture surface. The effect is particularly
well observable in the fatigue experiment. The repeated cycling
favors microcrack coalescence by longitudinal void propagation
along the boundary between adjacent microfibrils which enhances
the separation and visibility of fibrils in the final crack.

Although most of the experimental data on plastic deformation
of crystalline polymers were collected on polyethylene, polyoxymethylene, polypropylene and to a much smaller extent on nylon
one feels rather certain that the main aspects, particularly the
three stages mentioned above of the mechanism of deformation are
the same for all crystalline polymers. Future research will have
to elucidate the differences of detail caused by differences of
chemistry, geometry and tacticity of polymers involved which show
up already during solidification in a wide variety of crystal
lattice and ability to crystallize.

Plastic Deformation of Spherulitic Structure

As a consequence of the complicated spherulitic morphology of
crystalline polymer solids and the coexistence of the crystalline
and amorphous phase the stress and strain fields are extremely
inhomogeneous and anisotropic. The actual local strain in the
amorphous component is on the average higher and in the crystalline
component smaller than the macroscopic strain. In the composite

structure the crystal lamellae act as force transmitters and the
amorphous layers between them are the main contributors to the
strain. Hence in a very rough approximation the Leonard-Jones or
Morse type force field between adjacent macromolecular chain sections[1,2] describes fairly well the initial reversible stress-strain
relationship of a spherulitic polymer solid almost up to the yield
point, i.e., up to a true strain of about 10%.

But an irreversible plastic component is detectable already
at much lower strain, e.g., between 1 and 2% in polyethylene.[3]
The first mechanism of plastic deformation is lamella slip[4] followed by rotation of stacked lamellae from perpendicular towards
more parallel orientation with the tensile stress.[3] In the case
of polyethylene spherulites with the radial orientation of b-axis
the elastic and also the first stage of plastic deformation yield
a b-axis orientation in the tensile stress direction.[5] Such an
orientation is a purely geometric effect and does not involve any
plastic deformation of lamellae.

The forces on lamellae perpendicular to the main tensile stress
direction soon overcome the cohesive strength of the amorphous
layers and lead to physical separation of lamellae. The compressive component of the force field on the fold surfaces of the
lamellae parallel to the applied stress and the tensile component
in the lamella plane make the chains tilt thus reducing lamella
thickness. Both effects are well observable by small-angle X-ray
scattering.[6]

An early stage of plastic deformation of lamellae involves
twinning and phase transformation.[7,8] In single crystals of polyethylene one could demonstrate two different kinds of transformation of orthorhombic to monoclinic lattice if the elongational
strain was between 110 and 010 direction. If, however, the strain
had a direction between 100 and 110 twinning along the ($1\bar{1}0$)
diagonal occurred yielding a new orientation of the a-axis almost
perpendicular to the original orientation. In a bulk sample with
initially all orientations equally probable such phase transformation and twinning yield very soon a preferential orientation of
the a-axis perpendicular to the tension in good agreement with
experimental data.[5] Phase change and twinning, however, yield
only a modest plastic strain. A larger deformation, up to 100%
or even more, can be achieved by chain tilt and slip.[8]

The spherulite with the densely packed lamellae radiating
from the center is a rather rigid structure, much stronger than
the boundary layers between adjacent spherulites. Consequently
at an early stage of deformation the high strain concentration in
the boundary layers between adjacent spherulites between stacks
of lamellae inside the spherulites becomes sufficient for local

crack formation (Fig. 1). The concurrent fracture of lamellae protruding the crack leads to formation of microfibrils bridging the crack. In a sample with well developed, large spherulites the number of such microfibrils at the boundary of spherulites is too small for preserving the mechanical cohesion and the sample fails in a more or less brittle manner. The fracture surface, however, with the great many microfibrils pulled out of the spherulites exhibits a substantial amount of plastic deformation. At very low temperature this effect is drastically reduced. In such a case the fracture proceeds along the boundaries between adjacent lamellae and spherulites with very little microfibril formation. Such fracturing turned out to be an excellent method for electron microscopical studies of spherulitic and microspheru-litic structures.[10]

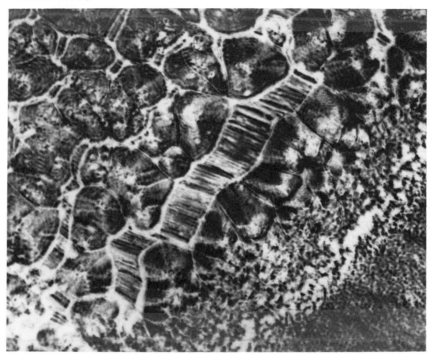

Fig. 1 – Optical micrograph of crack formation at the boundaries between adjacent spherulites of a thin linear polyethylene film deposited on a Mylar substrate. The cracks are bridged by many fibrils (Ingram[9]).

A special case is spherulites with helically twisted lamellae (Fig. 2). The geometry of the twist quite efficiently hampers the sliding motion of lamellae. In the equatorial plane of the spherulite the lamellae perpendicular to the stress are physically

Fig. 2 - Surface replica electron micrograph of a fracture surface
of a linear polyethylene banded spherulite, fractured in
liquid nitrogen showing the helical twisting of lamellae
in radial direction (Fischer[11]). At A the lamellae are
parallel to the fracture plane and will be physically
separated by a tensile stress perpendicular to that plane
and at B the lamellae are perpendicular to the plane and
will be laterally compressed and longitudinally extended
in the stress direction. In the latter areas the lamellae
will be soon fractured yielding bundles of microfibrils
while in the former areas this process must be preceded
by lamella rotation out of the perpendicular orientation
to the stress.

separated at an early stage of plastic deformation but seem to resist fracture for quite a while. In the lamellae parallel to the
stress, however, the chain tilt under frontal compression leads
rather soon to lamella rupture and crack formation. This alter-
nation of strikingly different deformational behavior with a radial
period equal to half the pitch of the helix shows up in the highly
anisometric onion-shaped layered structures one observes by optical
microscopy in drawn polymers at intermediate draw ratios.[12]

Transformation of Spherulitic into Fibrous Structure

Plastic deformation of unoriented more or less spherulitic
crystalline polymers during drawing with (cold drawing) or without
(hot drawing) a neck formation transforms the sample into a completely new highly oriented fibrous structure. During this transformation each folded chain lamella of the starting material is
broken and the crack bridged by a great many microfibrils if the
crack is at a non-vanishing angle to the growth plane so that it

cuts through a large number of chain folds.[13] The covalent forces
between consecutive chain atoms are so strong that very few chains
are broken by the crack as shown by the undetectably small number
of radicals formed during plastic deformation.[14] On the contrary,
the forces transmitted by the chains are sufficient for partial
chain unfolding, chain slip and for the breaking off of blocks
of folded chains which are then incorporated into the microfibrils.
The unfolded chains connect as tie molecules the blocks in axial
direction.

In order to understand better the microfibril formation one
has to investigate the effect on individual crystals.[7,8,15-18]
Usually one deposits them on a Mylar film or on a carbon coated
substrate which can be deformed some hundred per cent. If the
crystals adhere well on the substrate, they will be uniformly
deformed up to very high strain (~100% or even more) so that the
phase transformation, twinning, chain tilt and slip can be followed
up. In the case of inhomogeneous adherence the lamella sections
which are physically separated from the substrate tend to fail on
areas of maximum defect concentration thus relieving the stress on
the rest of the sections. The situation is still more extreme on
a carbon coated substrate. The brittle carbon layer cracks very
early so that the contingent section of the deposited lamella is
at an extreme strain which breaks the lamella while the rest remains practically undeformed. One has the feeling that in the
bulk sample one has both effects, a continuous plastic deformation
up to high strain in some lamellae and a very early fracture with
microfibril formation in other lamellae. All intermediate cases
are imaginable as a consequence of local variations of morphology,
lamella orientation and particularly interlamella adhesion.

Such experiments on polyethylene,[7,15] polypropylene,[16] polyoxymethylene[17] and nylon[18] single crystals demonstrated many
important aspects of microfibril formation. The transition from
the lamella to the microfibril is a microneck where the folded
chain blocks are torn away from the lamella, rotated by about 90°,
and incorporated in the microfibril. The connecting elements are
the partially unfolded chains at the boundary between adjacent
blocks which as tie molecules in the microfibrils are responsible
for its high longitudinal strength. The transformation from the
lamella to the microfibrils very likely involves some local preparation, chain tilt and extensive slip, development or isolation
of a mosaic structure[19,20] showing up in increased roughness of
fold surface, before the abrupt final step takes place in the microneck. The microfibrils do not vary enormously in thickness. The
electron micrographs show a tendency for coalescence of microfibrils which reduces the surface free energy. Very often a thick
microfibril originates from two distinct micronecks. Such a geo-

metry can be interpreted as a merging of two initially separated
microfibrils. Some micrographs,[21] however, show a nearly uniform
wide band or ribbon pulled out of the broken lamella which may
later be split into individual microfibrils. The experimental
data do not allow a clear decision between these two possibilities
although the observation of coalescent microfibrils seem to be more
frequent.

On the basis of the molecular model of folded chain crystals
one expects microfibril formation only in the case that the crack
is at an angle to the growth plane so that it cuts through a great
many folds. Such is the situation with polyethylene, polyoxymethylene, and nylon. In polypropylene the long lath-like single
crystals, however, break along the lath axis without and perpendicular to it with microfibril formation in spite of the fact that
the crystal growth seems to proceed by chain deposition on the 100
plane perpendicular to the lath axis.[16]

The length of the blocks in the microfibrils agrees fairly
well with the thickness of original lamella.[22] One would also
like to identify their lateral dimensions with the lateral dimensions of the mosaic blocks of the lamella. No systematic study of
the effect is available so that one has no experimental support for
or against such a guess.

In bulk samples which are cold drawn one can follow up by
electron microscopy the gradual transformation of spherulitic into
fibrous structure which occurs in the neck. Scanning electron
micrographs of the polyethylene neck (Fig. 3) show first in the
almost undeformed spherulitic material the formation of small
cracks more or less perpendicular to the tensile stress direction.
The cracks are bridged by fibrils. Proceeding farther in the
direction toward the fibrous structure the cracks grow thicker.
The micronecks at the ends of the fibrils gradually transform the
spherulitic material into fibrils so that at the end of the neck
almost all the sample exhibits the fibrous structure. Transmission
electron micrographs of surface replicas of polypropylene neck
(Fig. 4) with more or less randomly distributed small cracks very
clearly demonstrate the microfibril formation at a zone of micronecks which are gradually destroying a stack of parallel lamellae
of the spherulitic structure. Similar micrographs were obtained
from the polyoxymethylene neck.[24]

One tends to imagine that the micronecking at the crack of
the lamella is fairly independent of the geometry and flow pattern
of the neck and that the draw ratio at the microneck is a basic
property of the material which depends on crack and lamella orientation, lamella morphology, defect concentration, molecular weight,

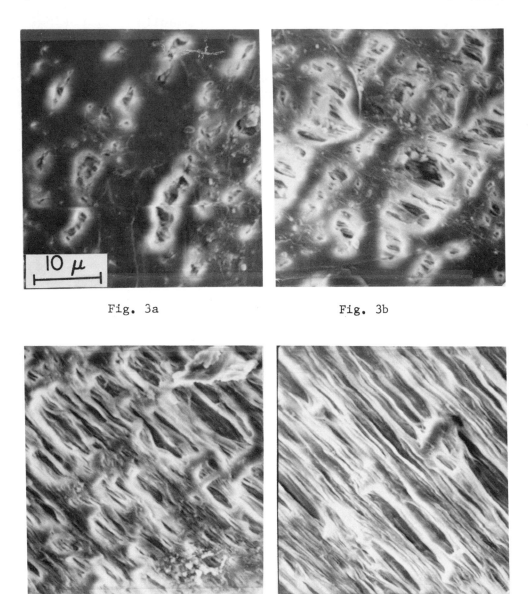

Fig. 3a   Fig. 3b

Fig. 3c   Fig. 3d

Fig. 3 - Scanning electron micrographs of progressive crack formation
         in the neck and transformation into fibrous structure.
         Linear polyethylene drawn at 60°C (Ingram[9]). In (a) at
         the beginning of the neck, the first cracks perpendicular
         to the draw direction become visible. They are bridged
         by fibrils. More advanced stages are shown in (b) and
         (c). At the end of the neck (d) almost all the spherulitic
         material is already transformed into fibrous structure.

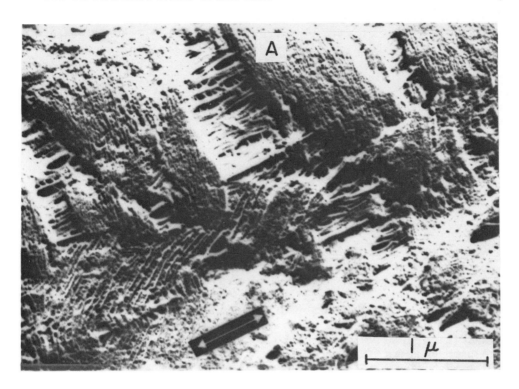

Fig. 4 - Surface replica electron micrograph of isotactic polypropylene film drawn at 65°C to a draw ratio 2 etched with
aqua regia at 0°C for 5 hours (Sakaoku-Peterlin[23]). The
cracks are bridged by a great many microfibrils (A) which
as a consequence of local variation of stress and strain
field are not completely aligned with the applied stress.
The still spherulitic material is showing some lamella
rotation, kinks, lamella slip and nearly undeformed stacks.

temperature, and draw rate. One hence expects a nearly constant,
time independent draw ratio in a drawing experiment. Contrary to
that the draw ratio of the volume element occurring during the
transformation from the spherulitic into fibrous structure is
small in the nascent neck which is a rather gentle constriction of
sample cross section and much larger in the mature neck with a
very sharp reduction of sample cross section. This conspicuous
difference may be a consequence of different draw ratio of microfibrils pulled out from the broken lamellae or more likely of a
different degree of structure transformation, small in the nascent
and nearly complete in the mature neck. The former alternative
seems rather unrealistic.

The coexistence of spherulitic and fibrous structure shows up in small- and wide-angle X-ray scattering,[25] optical birefringence, IR dichroism,[26] and density.[27] A particularly simple parameter for the estimate of the degree of transformation is the orientation of the crystal lattice which rapidly increases with the draw ratio. The spherulitic structure is very little and the fibrous structure is almost completely oriented. A great many analyses showing the close correlation between the properties of the drawn material and chain orientation[28] at a given temperature are merely proving the close correlation between mechanical properties and the fraction of fibrous material present which is indeed proportional to the observed chain orientation. If there is no fibrous structure, e.g., in an oriented single crystal mat, the orientation itself does not represent a good parameter for the description of the mechanical properties of the sample.

A more temperature dependent indicator is the orientation of the amorphous component which increases less rapidly than that of the crystal lattice.[26] It is indeed almost proportional to the fraction of fibrous structure if the drawing is performed at a temperature where the relaxation of the amorphous component is still negligible. Annealing or drawing at high temperature close to the melting point, however, reduces very rapidly the orientation of the amorphous material and much less that of the crystals so that the amorphous orientation ceases to be a reliable indicator of the fraction of fibrous material. But it seems to be very closely connected with the mechanical properties of the drawn material. Elastic modulus and stress-to-break decrease with annealing temperature and time in almost the same manner as the orientation of the amorphous component.[27]

With sufficiently high draw ratio the transformation into the fibrous structure seem to be completed. Electron micrographs of surface replicas convincingly demonstrate the well aligned microfibrils to be the basic structural element of the fibrous material.[23] The lateral autoadhesion of microfibrils is relatively poor. They can be easily separated from each other or sheared so that long longitudinal cracks or voids become visible (Fig. 5). The cracks are often bridged by microfibrils at oblique angles. Their orientation is a consequence of crack opening and shearing displacement of both sides of the crack against each other. One is faced with a similar situation in the longitudinal chopping of a piece of wood before complete separation. The gap is bridged by a great many wood fibers at angles which depend on the shearing displacement of the two sides and on the width of the crack as shown in Fig. 6 under the self-evident assumption that all the fibers are parallel in the wood log as grown.

Fig. 5 - Surface replica electron micrograph of highly drawn linear
polyethylene showing the microfibrillar structure and
longitudinal cracks bridged by more or less disoriented
microfibrils (Sakaoku[30]).

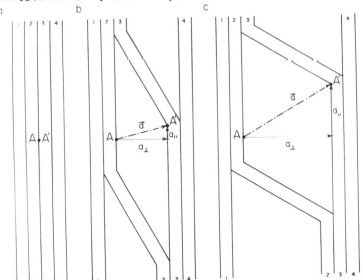

Fig. 6 - Model of fibrous structure showing the disorientation of
originally completely aligned microfibrils 2 and 3 (a)
after the longitudinal crack is formed (b, c) as a conse-
quence of lateral ($a_\perp$) and shear ($a_{||}$) displacement of
the crack boundary.

Surface replicas of annealed drawn polyethylene show the
existence of a fibrillar superstructure (Fig. 7). Bundles of
microfibrils with a lateral width of some thousand angstroms seem
to differ from adjacent bundles in draw ratio and hence in their
resistance to morphological changes during annealing. One concluded that each such bundle or fibril has originated from one
stack of parallel lamellae. Its transformation by a thin zone of
micronecks on the boundary of the lamellae into the fibril composed
from a great many microfibrils is schematically shown in Fig. 8.
As a consequence of varying angle between crack and lamella orientation from stack to stack one expects a corresponding variation
of the draw ratio of the resulting fibrils.

Fig. 7 – Surface replica electron micrograph of linear polyethylene
         drawn at 60°C to $\lambda = 8$ and subsequently annealed for 1
         hour at 110°C and etched with fuming nitric acid for
         10 hours at 80°C (Peterlin-Sakaoku[31]).

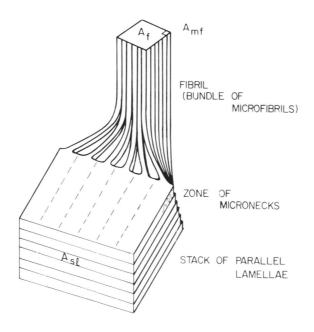

Fig. 8 - Model of transformation of a stack of parallel lamellae
by a thin zone of micronecks into a single fibril.

Long Period Change During Drawing

Small-angle X-ray investigation of the fibrous structure has
convincingly demonstrated the almost unique dependence of the long
period on the temperature of drawing and the complete independence
of the long period of the starting material.[32,33] This observation
is in maximum possible contrast with the drawing of single crystals
where the long period of the microfibrils is determined by the
thickness of lamella (Table I). Experiments at different draw ratio
have shown that the long period of the fibrous material depends on
the temperature established in the micronecking zone under adiabatic
conditions.[34] The heating above the temperature of the environment
is caused by the work of plastic deformation connected with the
destruction of lamellae.[35] The actual temperature rise in the
neck is less than that calculated under adiabatic conditions because
heat is lost to the environment by convection.[36] At sufficiently
low draw rate the heating is negligible indeed. But the local
heating in the destruction zone and in the small area behind it,
i.e., in a volume element with a radius of a few 1000Å, is adiabatic and not affected by heat exchange with the environment.[31]
The so calculated true temperature of drawing is so far below the
melting temperature (Fig. 9) that the observed adjustment of the
long period to the value corresponding to this temperature cannot

TABLE I

Long Period of Drawn Polyethylene as Function of Draw Temperature
T_d, Draw Ratio λ, and Long Period L_o of Starting Unoriented Material

	T_d	λ	L_o	L
Quenched Film[32]	24°C	11	195Å	165Å
	60°C	8	195	170
		10		174
		15		181
		20		176
	100°C	7		210
	120°C	10		255
Film[32] annealed 10^2 min at 120°C	60°C		250	180
10^2 125			310	170
10^2 125			380	180
Single crystals[22] from xylene solution	25°C		135	140
annealed at 127°C			160	180
127.5°C			230	240

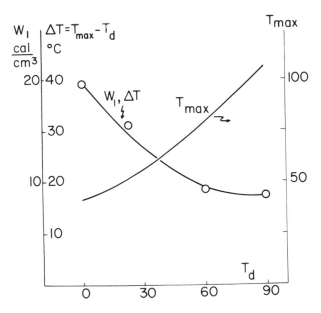

Fig. 9 – Adiabatic heating ΔT caused by the mechanical work W_1
during destruction of lamella and the maximum attained
temperature $T_{max} = T_d + \Delta T$ as a function of drawing
temperature T_d. Linear polyethylene, draw rate 0.5
cm/min (Meinel-Peterlin[35]).

be explained by local melting and recrystallization. Moreover,
the temperature is also below the value --110°C in the case of polyethylene -- where a remarkable long period growth is observable
during annealing.[37] The most striking long period adjustment during
drawing occurs with low temperature drawing of samples having a large
long period. In such a case the new long period is substantially
smaller than that of the starting material. Therefore, neither
melting and recrystallization nor long period growth as observed
during annealing can explain the observed change of long period
during drawing of bulk samples.

One may speculatively suggest the following explanation.[13]
The destruction zone transforms a certain cross section A of
lamellae into a cross section A/λ of the fibril with the draw
ratio λ. The substantial reduction of cross section produces a
high negative lateral pressure which tends to increase the inter-
chain distances in the crystal lattice. As a consequence the
ability of longitudinal translation of the chains is so much
increased that the whole crystal block can assume the length which
at the local temperature corresponds to the minimum of free
energy.[38-41] As soon as the negative lateral pressure ceases to
exist -- and that is expected to occur a short distance behind
the micronecking zone, may be at the same distance where the adiabatic heating gets modified by heat exchange with the environment
-- the crystal shape is fixed and does not undergo any farther
change by chain translation because at zero negative pressure the
temperature is too low for observable longitudinal chain mobility.

The effects do not play any role in the deformation of an
isolated single lamella because in this case the microfibrils are
individually free in air or vacuum and not encapsuled in the bulk
material which could cause a negative pressure as a consequence of
the reduced cross section. Therefore, the microfibril at the crack
is under no negative pressure which could in the crystal blocks
enhance the longitudinal chain mobility. Moreover, the linear
arrangement of micronecks at both sides of the lamella crack favor
rapid heat dissipation so that there is much less or even no
adiabatic heating of the microfibrils in contrast with the situation
in bulk samples. As a consequence of smaller temperature rise
and no lateral lattice expansion the folded chain blocks retain
their height.

Plastic Deformation of Fibrous Structure

Very little is known about this stage which plays an important
role at the end of the drawing process after the spherulitic structure has been completely transformed into the fibrous structure.
The load-elongation curve at constant elongation rate shows up to
the final fracture a steady increase of load above the value of

the almost constant drawing load. This phenomenon, often called
strain or work hardening, reflects the fact that the fibrous
structure has more than a λ-times higher resistance to plastic
deformation than the spherulitic starting material. In the true
stress-strain curve of the volume element (Fig. 10) one sees that
the hardening occurs immediately as the neck passes through the
element and continues at a much lower rate up to the fracture.[35,36]

The morphological explanation of the effect is the microfibrillar structure of fibrous material. The extremely thin ($\sim 100\text{Å}$)
and long ($\sim 10\mu$) microfibrils[42] are a very strong element as a
consequence of the great many almost completely extended tie molecules connecting subsequent crystal blocks. Their large number,
up to 30% of the chains in the crystal lattice act as tie molecules
in highly drawn polyethylene,[43] imparts to the microfibril a high
longitudinal strength and elastic modulus. Infrared investigations
have also demonstrated that the tie molecules in the drawn sample
are almost completely oriented and stretched thus forming rather
rigid bridges between subsequent crystalline blocks of the microfibril. They act almost as a crystalline connection between the
blocks thus extending the coherence of crystal lattice[44] and resisting the attack by fumic nitric acid many times better than the
fully relaxed amorphous material.[43] The high degree of orientation
of the amorphous material consisting of chain loops, free chain

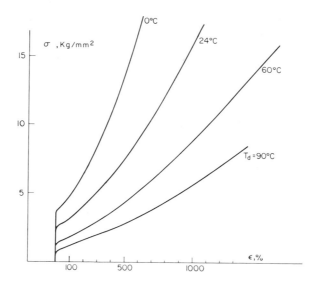

Fig. 10 - True stress-strain curves of linear polyethylene drawn
        between 0° and 90°C at a draw rate of 1 cm/min (Meinel-
        Peterlin[35]).

ends and tie molecules also drastically reduces the number of
sorption sites and the mobility of sorbent molecules as demonstrated
by the dramatic decrease of sorption, diffusion and permeability
of fibrous material[45] as compared with the transport properties
of undrawn samples.

    In a stress field the microfibril reacts very nearly as a
rigid structure much stronger than any other element of the polymer
solid. Since the boundary between adjacent microfibrils is not in
crystalline register one expects that the van der Waals forces
between adjacent microfibrils are weaker than those between adjacent
chains in the crystal lattice although stronger than in the amorphous
layer between adjacent lamellae of the spherulitic structure. From
this morphology one can conclude that the plastic deformation will
first involve the sliding motion of microfibrils which is opposed
by the friction resistance in the boundary. In spite of the weakness of the van der Waals forces the friction resistance over the
100 times larger length than the thickness of crystal lamella,
will be much larger than the forces opposing plastic deformation
in spherulitic material. Hence the strain hardening effect in
the fibrous material is a straightforward consequence of the large
surface to volume ratio of the strong microfibrils which efficiently
blocks shearing displacement of microfibrils and explains the increase of elastic modulus and stress to break by more than a decade
at sufficiently high draw ratio.

    The same morphology explains the angular dependence of plastic
deformation of fibrous material.[46,47] As far as possible, the
deformation occurs by sliding motion of microfibrils. If the new
draw direction is not exactly perpendicular to the fiber axis the
flow lines are parallel to the fiber axis yielding an asymmetrically
shaped sample. Deformation beyond yield occurs without any work
hardening and can be described in terms of flow behavior. Besides
microfibril sliding one also observes kink band formation[48,49]
which corroborates the basic aspects of the model of fibrous structure, particularly the prominence of microfibrillar strength and
the weakness of interfibrillar forces.

Acknowledgment

    This work was generously supported by funds from the Camille
and Henry Dreyfus Foundation. Thanks are due to Dr. P. Ingram for
the kind permission to include his optical micrographs of deformed
linear polyethylene membrane (Fig. 1) and his scanning electron
micrograph of the polyethylene neck (Fig. 3) and to Prof. Dr. K.
Sakaoku for the surface replica electron micrograph of drawn linear
polyethylene (Fig. 5).

References

1. N. Brown, Bull. Amer. Phys. Soc. 16, 428 (1971).
2. R. N. Haward and J. R. MacCallum, Polymer 12, 189 (1971).
3. R. S. Stein, Polymer Eng. Sci. 9, 172, 320 (1969).
4. P. Predecki and A. W. Thornton, J. Appl. Phys. 41, 4342 (1970).
5. S. L. Aggarwall and O. J. Sweeting, Chem. Revs. 57, 665 (1956).
6. F. J. Baltá-Calleja and A. Peterlin, J. Macromol. Sci. B4, 519 (1970).
7. P. H. Geil, J. Polymer Sci. A-2, 3813 (1964).
8. H. Kiho, A. Peterlin, and P. H. Geil, J. Appl. Phys. 35, 1599 (1964).
9. P. Ingram, unpublished work.
10. See for instance P. H. Geil, Polymer Single Crystals, J. Wiley & Sons, New York, 1963.
11. E. W. Fischer, Z. Naturf. 12a, 753 (1957).
12. J. D. Muzzi and D. Hansen, Text. Res. J. 41, 436 (1971).
13. A. Peterlin, J. Material Sci. 6, 490 (1971).
14. D. Campbell and A. Peterlin, J. Polymer Sci. B6, 481 (1968).
15. A. Peterlin, P. Ingram, and H. Kiho, Makromol. Chem. 86, 294 (1965).
16. P. Cerra, D. R. Morrow, and J. A. Sauer, J. Macromol. Sci. B3, 33 (1969).
17. K. O'Leary and P. H. Geil, J. Macromol. Sci. B2, 261 (1968).
18. K. Sakaoku and A. Peterlin, Makromol. Chem. (in press).
19. C. A. Garber and P. H. Geil, Makromol. Chem. 113, 251 (1968).
20. G. S. Y. Yeh and P. H. Geil, J. Macromol. Sci. B2, 29 (1968).
21. P. H. Geil, Polymer Single Crystals, J. Wiley & Sons, New York, Fig. VII.37, p. 451.
22. P. Ingram, Makromol. Chem. 108, 281 (1967).
23. K. Sakaoku and A. Peterlin, J. Polymer Sci. A-2, 9, 895 (1971).
24. A. Siegmann and P. H. Geil, J. Macromol. Sci. B4, 557 (1970).
25. A. Peterlin and F. J. Baltá-Calleja, Kolloid-Z. & Z. Polymere 242, 1093 (1970).
26. W. Glenz and A. Peterlin, J. Macromol. Sci. B4, 473 (1970); J. Polymer Sci. A-2, 9, 1191 (1971); Makromol. Chem. 150, 163 (1971); Kolloid-Z. & Z. Polymere 247, 786 (1971).
27. W. Glenz, N. Morosoff, and A. Peterlin, J. Polymer Sci. B9, 211 (1971).
28. R. J. Samuels, J. Polymer Sci. A-2, 6, 1101, 2021 (1968); 7, 1197 (1969); J. Macromol. Sci. B4, 701 (1970).
29. G. Meinel and A. Peterlin, J. Polymer Sci. B5, 613 (1967).
30. K. Sakaoku, unpublished work.
31. A. Peterlin and K. Sakaoku, J. Appl. Phys. 38, 4152 (1967).
32. R. Corneliussen and A. Peterlin, Makromol. Chem. 105, 193 (1967).
33. F. J. Baltá-Calleja and A. Peterlin, J. Materials Sci. 4, 722 (1969).
34. F. J. Baltá-Calleja, A. Peterlin, and B. Crist, J. Polymer Sci. A-2, 10, 00 (1972).

35. G. Meinel and A. Peterlin, J. Polymer Sci. A-2, $\underline{9}$, 67 (1970).
36. P. Vincent, Polymer $\underline{1}$, 7 (1960).
37. E. W. Fischer and G. F. Schmidt, Angew. Chem. $\underline{74}$, 551 (1962).
38. A. Peterlin, J. Appl. Phys. $\underline{31}$, 1934 (1960).
39. A. Peterlin and E. W. Fischer, Z. Phys. $\underline{159}$, 272 (1960).
40. A. Peterlin, E. W. Fischer, and C. Reinhold, J. Chem. Phys. $\underline{37}$, 1403 (1962).
41. A. Peterlin and C. Reinhold, J. Polymer Sci. $\underline{A-2}$, 2801 (1965).
42. A. Peterlin, Text. Res. J. $\underline{42}$, 20 (1972).
43. G. Meinel and A. Peterlin, J. Polymer Sci. A-2, $\underline{6}$, 587 (1968).
44. G. Meinel, N. Morosoff, and A. Peterlin, J. Polymer Sci. A-2, $\underline{8}$, 1723 (1970).
45. J. L. Williams and A. Peterlin, J. Polymer Sci. A-2, $\underline{9}$, 1483 (1971).
46. Y. Tsunekawa, M. Oyane, and K. Kojima, J. Polymer Sci. $\underline{50}$, 35 (1961).
47. T. Hinton and J. G. Rider, J. Appl. Phys. $\underline{39}$, 4932 (1968); J. Materials Sci. $\underline{6}$, 558 (1971).
48. R. E. Robertson, J. Polymer Sci. A-2, $\underline{7}$, 1315 (1969); $\underline{9}$, 453 1255 (1971).
49. K. Imada, T. Yamamoto, K. Shigematsu, and M. Takayanagi, J. Materials Sci. $\underline{6}$, 537 (1971).

ORDER IN ATACTIC POLY(PARA-BIPHENYL ACRYLATE) AND RELATED POLYMERS

B. A. Newman, Rutgers University, N.J., U.S.A.

P. L. Magagnini and V. Frosini, University of Pisa, Pisa, Italy

ABSTRACT

The unusual properties of atactic poly(parabiphenyl acrylate
reveal that some degree of order in this polymer exists. An x-ray
study of this polymer and some related polymers confirm this and
indicate that the order is one-dimensional. Some annealing studies
show that the extent of this order can be increased by thermal
treatments. A model of the molecular packing is suggested to
account for these results.

Introduction

The production of sharp x-ray diffracted intensities from
a polymer by one of the conventional x-ray techniques is usually
taken as indicative of crystallinity in the polymer sample. However, in as much as a crystal is defined as a regular spatial re
peat in three dimensions of some atomic configuration over some
long range distance, it should be observed that in certain cases
sharp diffraction can also occur from spatial repeats in one or
two dimensions. Materials with such an order are then not
crystalline in the conventional sense and can be spoken of as
possessing one or two degrees of order. This phenomenon is well
known to occur in liquid crystals where both one and two dimensional degrees of order are found. The long spacing observed in
many crystalline polymers is indicative of a one-dimensional
super-order imposed on the normal three dimensional order.
Following a study of many different types of polymers showing one
and two-dimensional degrees of order Statton generalized his results by concluding that two-dimensional degrees of order could
be attributed to a regular lateral packing of linear molecules
with no order along the chain direction (longitudinal disorder)
and one-dimensional degree of order to a longitudinal periodicity
with no lateral order. Longitudinal disorder could arise from
atacticity or a composite nature of the repeating unit. Lateral
disorder could be due to non-linearity of the chain or local dipole repulsion, the longitudinal periodicity arising from the
chemical repeat along the chain. Several examples confirm these

ideas, poly(trimethylene terephthalamide) showing only longitudinal order polyacrylonitrile[2], poly(vinyl trifluoroactetate)[2] showing only lateral order. Another example was provided more recently by poly-N-vinylcarbazole[3], an x-ray study showing that the molecules were stiff rod-like molecules which packed well laterally but with no longitudinal regularity. Recently a report on the preparation and characterization of poly(para-biphenyl acrylate) PPBA[4] was made. Although the polymer as prepared was atactic a number of experimental observations indicated that the polymer had some characteristics of a crystalline polymer. In this study PPBA and some related polymers were studies with x-ray methods.

Experimental

The polymers investigated were:

poly(para-biphenyl acrylate)PPBA

$\sim CH_2-CH-CO-O-\hexagon-\hexagon$

poly(para-cyclohexyl phenyl acrylate) PPCPA

$\sim CH_2-CH-CO-O-\hexagon-\hexagon H$

and poly(para-biphenyl methacrylate) PPBMA

$\sim CH_2-C(CH_3)-CO-O-\hexagon-\hexagon$

PPBA was observed to show a distinct melting over a range 255-260°C and PPCPA a melting range 190-195°C. PPBMA did not show any distinct melting range. The samples were studies at room temperature using a Norelco wide-angle x-ray diffractomater using CuKα radiation. X-ray scans revealed a single sharp peak for PPBA and PPCPA at Bragg angles of ∿4.6° respectively. No sharp peak could be observed in the case of PPBMA, although a general very broad peak, which was obviously diffraction from amorphous material was observed around $2\theta \sim 15°$.

Annealing experiments were made to attempt to increase the extent of order present in these polymers. The results are shown in Table I and figures (I), and (II).

The conditions of measurement with the diffractometer and the disposition of the sample were maintained as constant as possible during the measurements, so that comparisons would be legitimate. In Table I the peak heights are shown following each annealing treatment for purposes of comparison. Annealing of PPBA for six hours at 245°C more than doubled the peak. However further annealing led to a decrease in the peak height. Annealing of PPCPA at 170°C for six hours approximately doubled the peak height and again further annealing led to a decrease. We attribute the decreased peak height intensities after extensive annealing as arising from degradation.

Table I

Sample	Annealing Temperature		Annealing Time (hours)	Peak Height (arbitrary units)
PPBA	unannealed	a		13
	245°C		3	15
	245°C	c	6	31
	245°C	b	9	19
PPCPA	unannealed	a		8
	170°C		3	10
	170°C	c	6	15
	170°C	b	9	13

Figure I – PPCPA

Figure II – PPBA

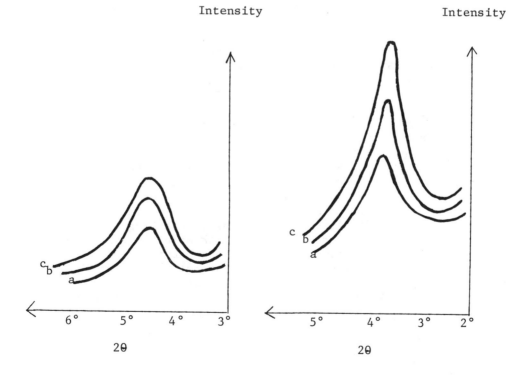

The polymer PPBMA showed no indication of order with any of the
annealing treatments tried.

Attempts were made to orient the samples by drawing. However
the glass transition temperatures of these polymers are above 100°C.
At room temperature they are extremely brittle. Tensile experiments
above 150°C usually resulted in fracture without orientation.
Using CuKα monochromatic radiation and a Guinier camera, very weak
2nd and 3rd orders of the principal reflections were observed
and none others, after very long exposure times.

Discussion

Some investigations of PPBA by Magagnini et al.[4] indicate that
this polymer possesses some very unusual properties indicative
of a high degree of order. The solubility of PPBA in benzene and
other solvents was found to be very much less than that of comparable polymers such as PPBMA. The density of PPBA (1.1236) is
appreciably higher than that of polymers of a similar structure.
The thermal behavior is unusual in that melting and solidification processes are easily observed by the sharp variation in optical
properties of the sample. Upon heating under pressure, the pressure
remained constant up to temperatures close to the melting range
and then within a few degrees the polymer changed to a viscous
liquid. Thus it was established that the solid-liquid transition
in PPBA is sharp, and that PPBA retains the properties of a solid
to a temperature close to this melting range. Moreover the
melting range (250-285°C) depending on sample history is unusually
high. In comparison PPBMA has a softening point below 160°C.
Since, with the polymerization method used, PPBA should be substantially atactic, one would expect a completely amorphous polymer
with a glass transition temperature and softening point between
100°C and 130°C.

The results of the x-ray diffraction study show that PPBA and
PPCPA possess a high degree of order, yet are not crystalline in
the conventional sense. Since these polymers are atactic it
was at first supposed that the order must be a two-dimensional
lateral packing with longitudinal disorder. The molecule cannot
however possess a cylindrical rod-like configuration and pack in
the same way as poly-N-vinyl carbazole and other reported examples
since calculation of the density shows that this would be too
loose a structure. Spacings corresponding to the Bragg reflections
at 3.8° and 4.6° for CuKα radiation are 23.2 Å and 19.2 Å and these
are much larger than is commonly found among laterally ordered
polymers. For poly-N-vinyl carbazole a lateral spacing of 10.7Å
was reported. Moreover such a model would not fit the x-ray
data. A two-dimensional lateral packing of rod-like molecules

with cylindrical symmetry would yield a pseudo-hexagonal set of
reflections. This was not observed. Only 2nd and 3rd order reflections of the principal spacing could be observed in each case
and only with difficulty. This model then was abandoned.

     The x-ray data indicate only a one-dimensional degree of order,
with a spacing of 23.2 Å for PPBA and 19.2 Å for PPCPA. It should
be noted that the side-chain branches are quite long (approximately
10Å) and also rigid, and approximately perpendicular to the main
chain. If we assume that statistically a pseudo-syndiotactic
conformation is adopted, the molecule would assume a ribbon-like
conformation, the width of the ribbon being approximately 20Å,
Fig. (III) shows this schematically.

Figure III

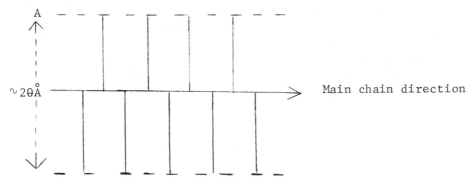

During polymerization one would expect from the kinetics of polymerization a syndiotactic conformation to be favored since it would
be energetically unfavorable to assume an isotactic conformation.
Since it would be most unfavorably sterically for such two long
side chains to be adjacent it is also possible that main chain
rotation and distortion might force a pseudo-syndiotactic conformation. For either of these reasons we assume the molecules to be
ribbon-like. It is to be noted that these considerations would not
apply with the same force for the case of PPBMA. During solidification a lateral packing of these ribbons would then form ∿20Å thick
sheets, figure IV, which would then account for the x-ray data.

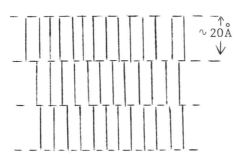

Figure IV

Main chain perpendicular
to figure

A completely syndiotactic conformation might be expected to be quite crystalline in the conventional sense. No other order is observed however and so the polymers PPBA and PPCPA must be dis- ordered within the $\sim$ 20Å sheets. Presumably the molecules are not sufficiently coplanar to allow a regular lateral packing within the sheets.

Very little difference in the diffraction patterns of PPBA and PPCPA was detected except for the smaller spacing for the latter polymer. The cyclo hexyl group should fail to give the resonance stabilized coplanar conformation of the two phenyl groups expected in PPBA. Also the cyclo hexyl groups are flexible and no-coplanar contrary to the phenyl group. These observations may account in part for the smaller spacing observed in the case of PPCPA. It is suggested that a smaller degree of syndiotacticity in the case of PPBMA would account for the feature that this polymer appears com- pletely amorphous. However, a number of other related polymers besides PPBMA such as poly-p-phenyl styrene, polyvinyl-p-phenyl- benzoate, poly-p-phenyl-benzyl acrylate are found to be amorphous and hence this explanation is not as yet complete.

It is intended to confirm the model suggested by an inves- tigation of oriented samples. Since the degrees of order obtained in these polymers arise from a delicate balance of steric and polar factors, additional study of other related polymers is intended.

References

1. W. O. Statton, Ann. N.Y. Acad. Sci. 83, 27, (1959).

2. C. R. Bohn, J. R. Schaefgen, and W. O. Statton, J. Pol. Sci., 55, 531 (1961).

3. A. Kinura, S. Yoshimoto, Y. Akana, H. Hirata, S. Kusabayashi, H. Mikawa, and N. Kasai, J. Pol. Sci. A-2, 8, 643 (1970).

4. M. Baccaredda, P. L. Magagnini, G. Pizzirani and P. Guisti, Pol. Letters, 9, 303 (1971).

A STUDY OF STRUCTURE OF COPOLYMER CRYSTALS: CHLORINATED

POLYETHYLENE CRYSTALLIZED FROM DILUTE SOLUTION

Ryong-Joon Roe and H. F. Cole

Bell Telephone Laboratories

Murray Hill, New Jersey   07974

and

D. R. Morrow

Rutgers University

New Brunswick, New Jersey

ABSTRACT

As a model of random copolymers, chlorinated polyethlenes (0.2 to 2 mole % Cl) were prepared by chlorination of linear polyethylene in solution. Crystallization of these from dilute xylene solutions results in aggregates of "single crystals." Morphology of these crystals were examined with an electron microscope. Although the regularity of the crystals deteriorated with increasing chlorine concentration, the lamellar nature of the crystal structure was still retained. Thermodynamic properties of these crystals prepared at different crystallization temperatures were analyzed to obtain information on their structure. The observed lamellar thickness as a function of crystallization temperature is found to obey the Hoffman-Lauritzen formulation originally proposed for homopolymers, but the kinetic surface free energy values deduced decreased with increasing chlorine contents. The equilibrium surface free energy values, obtained from melting temperature dependence on lamellar thickness, also are found to decrease similarly and agree approximately with the kinetic surface free energy values.

I. INTRODUCTION

Crystallization kinetics of homopolymers and the structure
and morphology of resultant homopolymer crystals have been studied
intensively by many in the past decade or two, and are now fairly
well understood. In comparison, much less is known about the properties and structure of semicrystalline copolymers. Theoretical
concepts and experimental techniques developed with respect to
homopolymers could now be employed to the study of copolymers with
fruition. From a practical point of view, the possibility of finding new polymers are diminishing, and one may instead seek subtle
modification of properties of existing polymers by copolymerization
and other means. In this context it is appropriate that we devote
more effort to the study of structure of copolymers.

In this work we employ a series of chlorinated polyethylenes
as a model compound for the study of random copolymers. They were
prepared by chlorination of linear polyethylene in solution. The
degree of chlorination in our samples were kept fairly low and at
most to a few chlorine atoms per 100 carbons. Chlorinated polyethylene offers several advantages over other copolymers for our
purpose. First, the chlorination reaction conducted in solution
results in the placement of chlorine atoms at random positions
along the polyethylene chain. This aspect is important in the
interpretation of properties of copolymers in term of their chemical structure. Direct copolymerization of mixed monomers, especially when conducted on a commercial scale, results frequently in
non-random copolymers. Secondly, the chlorination reaction is easy
to perform on a laboratory scale, and the degree of chlorination
can be controlled readily. Thirdly, the atomic number of chlorine
is sufficiently different from carbon and hydrogen, and the resultant high contrast in both mass and electron density can facilitate
the use of physical techniques, such as x-ray diffraction, in the
study of crystalline structure of these material.

In this work a study is made of the crystallization of these
chlorinated polyethylenes from dilute solution, and also of the
thermodynamic properties of the resulting single crystals. In
analyzing the results, we have paid attention to finding the extent
of similarity between the behavior of copolymers with that of homopolymer. In separate work the structure of the same copolymer
crystallized from melt, instead of dilute solutions, was studied
by the technique of small angle x-ray diffraction and the result
will be published shortly.[1] In addition, a comparative study of
dielectric behavior of these chlorinated polyethylene with that of
linear polyethylene was undertaken previously and was published
elsewhere.[2]

II. EXPERIMENTAL

Chlorinated polyethylene samples were prepared and kindly supplied to us by E. P. Otocka of this laboratory. They were prepared by passing chlorine gas through a solution of Marlex 6050 in tetrachloroethylene (at 105°C). Details of the preparation method are described elsewhere.[3] Compositions of the samples are tabulated in Table 1. The control sample designated No. O is Marlex 6050 which has gone through the process of dissolution in tetrachloroethylene and reprecipitation without chlorination.

Single crystals were prepared by isothermal crystallization mostly from a dilute solution containing 0.04g of polymer per liter of xylene. Some attempts were also made to grow crystals from seeds suspended in the dilute solution by a procedure similar to that described by Blundell et al.[4] for preparation of single crystals of polyethylene. The technique appeared, however, to grow less effective as a means of inducing crystallization in general as the amount of chlorine substitution increased. The crystals obtained with or without seeding were collected by filtration and washed repeatedly with fresh solvent at the crystallization temperature. In some instances, the final washing was done with acetone or benzene to facilitate later drying of the crystal mat.

Since not all the copolymer initially dissolved is collected as crystals, there arises the possibility that a partial fractionation accompanies crystallization and the chlorine content in the collected crystals becomes different from that in the original material. In a series of experiments, therefore, the concentrations of chlorine in the single crystals as well as in the residue reclaimed from the mother liquor were determined by a wet chemical analysis. The results are tabulated in Table 2. In the first series of the experiments, single crystals of the same copolymer

Table 1

CHLORINATED POLYETHYLENE SAMPLES

Sample No.	Wt.% of Chlorine	No. of Cl atoms per 100 cartons	Melting Temperature T_m
No. 0	0	0	419° K
No. 1	0.49	0.19	416
No. 2	1.29	0.49	413
No. 3	3.29	1.34	406
No. 4	4.20	1.76	--

Table 2

CHANGES IN CHLORINE CONCENTRATION IN CRYSTALS
GROWN FROM DILUTE SOLUTION

Sample	Cl Content (wt.%)
Starting Material	3.51
Single Crystals, T_c=77°C.	3.28
Single Crystals, T_c=73°C.	3.33
Single Crystals, T_c=65°C.	3.24
Residue in the mother liquor	5.18
Starting Material	4.02
Single crystals, T_c=65°C.	3.70
Residue in the mother liquor	5.18

No. 3 were grown at different temperatures of crystallization T_c, but the chlorine concentration remains about the same and is only slightly lower than that of the starting material. The same is true from the second series of the experiments in which the starting copolymer No. 4 contains higher concentration of chlorine. It appears fair to say that the degree of fractionation on crystallization from solution is very small, and the amount of chlorine in the single crystals can still be approximated by that of the original copolymers.

Melting temperatures were determined by use of a differential scanning calorimeter, Perkin-Elmer, Model 1B. The scan rate used was 10°C/mins. It is known that the determination of melting temperature of polyethylene single crystals is often made difficult by structural reorganization of these crystals during heating which results in multiple peaks in the thermogram. This difficulty is frequently overcome[5] by light irradiation of crystals which introduces a small amount of crosslinks hindering reorganization. With materials of high chlorine concentration (1.34 or more chlorine atoms per 100 carbons), no multiple peaks were observed, which probably indicates that chlorine atoms are also effective in hindering reorganization of crystalline structure. With crystals of lower chlorine contents, light irradiation was found necessary to suppress the formation of multiple peaks in the thermogram. More detailed description of reorganization behavior of these crystals is given elsewhere.[6]

Lamellar thickness of single crystal mats was determined from a low-angle x-ray photograph obtained with a camera similar to that described by Statton.[7] Intensity maximum in the low-angle region was observed with all the single crystals, and a second order diffraction peak was also observed with most of the specimens. The

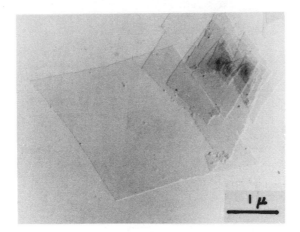

Figure 1.   Crystals from the copolymer containing 0.49 chlorine
per 100 carbon atoms, grown in xylene at an undercooling of 23°C
$(T_c = 86°C)$.

crystalline unit cell dimensions were determined from x-ray diffraction photographs obtained with a Guinier-deWolff camera. Diffraction lines from sodium chloride crystals adhered to the surface
of copolymer samples were recorded on the same photograph and
served as a calibration standard. All the diffraction lines from
the polyethylene crystalline structure (up to 22 lines) were employed
in the least square determination of the orthorhombic dimensions a,
b, and c.

III. MORPHOLOGY OF SINGLE CRYSTALS

     Electron-microscopic observation of crystals obtained from
xylene solution indicates that the regularity and appearance of
crystals deteriorate with increasing chlorine contents, but formation of a monolayer lamellar structure could be discerned with
materials containing up to 4.33 chlorine per 100 carbon atoms.
Crystals of a copolymer containing 0.19 chlorine per 100 carbons are
virtually identical in appearance to those obtained from linear
polyethylene. When the chlorine content is increased to 0.49 chlorine per 100 carbons, slight irregularity introduced into the single
crystals becomes discernable. Figure 1 is an electron micrograph
of such crystals grown at a supercooling of 23°C (temperature of
crystallization T_c = 86°C). With further increase in the chlorine
content, the crystal shape becomes noticably rounded. Figure 2
illustrates crystals of a copolymer containing 1.34 chlorines per
100 carbons and grown in xylene at a supercooling of 24°C(T_c = 78°C).
With this copolymer the word "single" crystals is very much appropriate even when they were grown at a much higher supercooling.

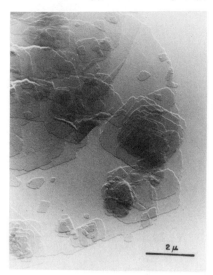

Figure 2.   Crystals of the copolymer containing 1.34 chlorine per 100 carbon atoms, grown in xylene at an undercooling of 24°C (T_c=78 ℃).

Figure 3.   A selected crystal of the copolymer containing 1.34 chlorine per 100 carbon atoms, grown in xylene at an undercooling of 28°C (T_c=74 ℃).

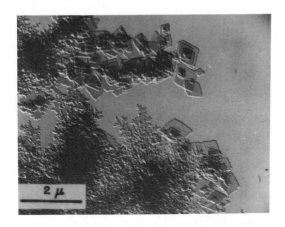

Figure 4. Crystals of the same copolymer at an undercooling of
56°C (T_c=46°C).

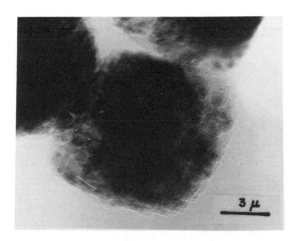

Figure 5. A highly overgrown crystal of the copolymer containing
1.74 chlorine per 100 carbon atoms, grown in xylene at a low undercooling (T_c=71°C).

Figures 3 and 4 show selections of crystals of this copolymer obtained at a supercooling of 28°C and 56°C (or T_c of 74°C and 46°C)
respectively.

The next copolymer, containing 1.74 chlorine per 100 carbon
atoms, gave a collection of highly overgrown crystals from which
it was difficult to identify many real "single" crystals. Figure
5 is a representative field from a batch of these crystals grown
at the highest practicable crystallization temperature, 71°C. The
basic building block in these complex structures is still a diamond
shaped flat platelet; this was confirmed by studying micrographs
obtained at higher magnifications, but could also be inferred from
the shapes of layers at the edge of the complex crystals shown
in Fig. 5. Figure 6 is an electron micrograph of a copolymer of
the highest chlorine content available (4.33 Cl/100 C). The crystal is highly overgrown, with no readily visible platelet structures.
But, on careful examination, it is still possible to find crystal
aggregates, such as shown in Fig. 7 which betray the lamellar
nature of the underlying structure.

IV.   THERMODYNAMICS OF GROWTH AND MELTING OF SINGLE CRYSTALS

The melting temperatures of the copolymers determined by DSC
are shown in Fig. 8 against the reciprocal of lamellar thickness ℓ.
The data for linear polyethylene shown are taken from the work of
Bair, Huseby and Salovey[5] for comparison. The melting points indi-

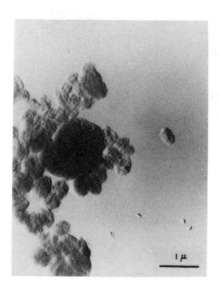

Figure 6. Aggregates of crystals of the copolymer containg 4.33
chlorine per 100 carbon atoms.

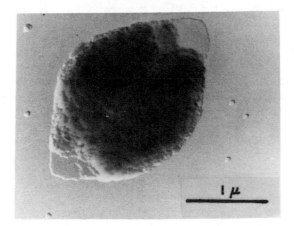

Figure 7. A selected example of highly overgrown crystals from the aggregates shown in Fig. 6.

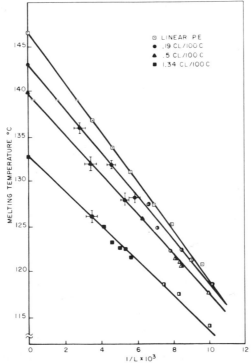

Figure 8. Melting temperatures of copolymer crystals plotted against the reciprocal of the lamellar thickness determined by low-angle X-ray technique. See the text for the explanation of symbols.

cated by half-filled symbols in Fig. 8 were obtained with single
crystal mats crystalilzed from xylene solution at different temperatures. The data shown with filled symbols were obtained with
crystals annealed at temperatures below their respective melting
points. The symbols with crosses represent data obtained with
materials crystallized from melt and thus of lower accuracy. The
sets of data points for each copolymer can be represented fairly
well by straight lines, in accordance with the relation

$$T_m = T_m^o + \frac{2\sigma_e T_m^o}{\Delta H_f \ell} \tag{1}$$

where σ_e is the surface free energy. The "equilibrium" melting
temperature T_m^o, obtained by extrapolation of T_m to $1/\ell = 0$, is
given in Table 1. From the slope of the straight lines, one can
calculate the ratio of $\sigma_e/\Delta H_f$, and hence the value of σ_e when
one knows the value of ΔH_f. This point will be discussed further
in a later paragraph.

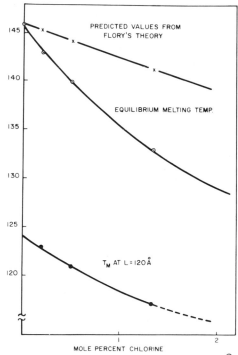

Figure 9. The equilibrium melting temperature T_m^o, determined by
extrapolation in Fig. 8, is plotted against the chlorine concentration in the copolymers. For comparison, the observed melting
temperature T_m of crystals of lamellar thickness equal to 120 Å,
and the theorectical value predicted from Flory's copolymer theory
are also given.

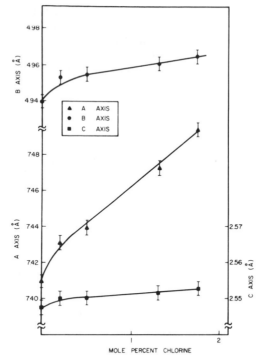

Figure 10. Unit cell dimensions of copolymer crystals, as determined by wide-angle X-ray diffraction, are plotted against the chlorine concentration.

The extrapolated melting temperature T_m^o is plotted against the concentration of chlorine in Fig. 9. Also shown there for the purpose of comparison are the values of T_m^o predicted by Flory's copolymer theory[8], i.e.,

$$1/T_m^o = 1/T_m^{oo} + (R/\Delta H_f) \ln X_A \qquad (2)$$

where T_m^{oo} is the melting point of homopolymer (extrapolated to $1/\ell = 0$), ΔH_f^o the heat of fusion of homopolymer and X_A the mole fraction of crystallizable units ($-CH_2-$). In Fig. 9 it is evident that the effect of increasing chlorine content is to depress the melting point of copolymers much more rapidly than is predicted by equation (2). Such discrepancy has been observed also with a number of other copolymer series[9-10]. In a separate study[1] on the same chlorinated polyethylene samples but crystallized from melt, we employed the technique of low angle x-ray scattering and reached the conclusion that some of the chlorine atoms actually become incorporated in the crystalline phases. It is probable that, even when the copolymers are crystallized from solution, the same tendency

for the chlorine atoms to enter crystalline phases remains. This
can be inferred also from the fact that the unit cell dimension
expands as the concentration of chlorine increases in the copolymers,
as is shown in Fig. 10. The theory by Flory, leading to Equation
(2) is an equilibrium theory in the sense that it assumes the
formation of ideal crystalline lattices from which all non-crys-
tallizable impurities are rejected. It is then not surprising
that the observed melting temperature does not in reality follow
the prediction of equation (2), as illustrated in Fig. 9.

The lamellar thickness ℓ of single crystal mats obtained
from dilute solution depends on the temperature of crystallization
in accordance with the relation

$$\ell = \frac{2\sigma_e T_d^0}{\Delta H_f \Delta T} + \delta \ell \tag{3}$$

where T_d^0 is the temperature of dissolution of the crystals in the
given solvent, extropolated to $1/\ell = 0$ and ΔT is the difference
between T_d^0 and the temperature of crystallization. The kinetic
theories[11] of formation of folded-chain homopolymer crystals all

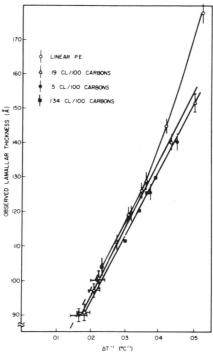

Figure 11. The lamellar thickness, determined by low-angle X-ray
technique, is plotted against the reciprocal of the supercooling
ΔT at which the crystals are grown in xylene.

Table 3

SURFACE FREE ENERGY

$(ergs/cm^2)$

Sample No.	Surface Free Energy	
	From Melting Point Data	From Growth Data
No. 0	89±5	112±8
No. 1	83±9	70±9
No. 2	77±10	70±9
No. 3	69±10	70±9

lead to equations essentially similar to (3), and the detailed differences among various theories are reflected in the term $\delta\ell$, which depends only moderately on temperature. According to Mandelkern[9] the basic assumption common to all nucleation theories would lead to the expectation that the size of crystal is related to the supercooling ΔT as indicated by the first term of equation (3). Figure 11 in which the observed lamellar thickness ℓ is plotted against $1/\Delta T$, indeed bears out this expectation. (Note: Computation of T requires the knowledge of the equilibrium disso-lution temperature T_d^o, which cannot be determined readily. We have therefore assumed that T_d^o in xylene is lower than T_m^o by 3P C for all the copolymers shown.)

From the slope of the plots in Fig. 11, one can obtain the ratio of $2\sigma_e/\Delta H_f$, according to equation (3), We now have two independent ways of determining σ_e values, one from the plot of T_m against $1/\ell$, and the other from the plot of ℓ against $1/\Delta T$. The values of σ_e obtained by these two methods are tabulated in Table 3. In these calculations, the value of H_f is taken to be 68 cal/g which is applicable to a perfect polyethylene lattice. The surface free energy values evaluated by the two methods do not necessarily refer to the same physical quantity. The value obtained from the melting point data relates to the specific free energy of the fold surfaces which is more nearly in the state of equilibrium as a result of annealing prior to melting, while the value obtained from the growth data refers to the surfaces which are freshly formed during the crystallization process. The distinction between the "equilibrium" vs "kinetic" surface free energies has been stressed by Hoffman[12,13] and his coworkers. The data in Table 3 indicates that these two quantities appear to be substantially the same and any difference observed between them is barely beyond the experi-mental error. We also note that the surface free energy appears to decrease moderately with increasing concentration of chlorine. This is apparently in contradiction to the common notion that the

surface tension of polar liquid is higher than that of non-polar
liquid. But the analogy with the surface tension of liquids is a
misleading one, since the major contribution to the fold surface
free energy σ_e arises partly from the work of forming chain folds,
and partly from the fact that there is a disordered layer of a
considerable thickness on the surface of crystalline lattice.[14]

In summary, in this work we employed a series of chlorinated
polyethylene as model compounds for the study of copolymers. Single
crystals of these chlorinated polyethylenes obtained from dilute
xylene solutions were examined with respect to their morphology and
thermodynamic properties. Experimental techniques and theoretical
concepts developed in the study of homopolymers were adapted. For
the low degrees of the comonomer content employed here, features
observed with these copolymers were qualitatively similar to those
previously found with homopolymers.

ACKNOWLEDGMENT

The authors wish to acknowledge Dr. E. P. Otocka for kindly
supplying the chlorinated samples, Mr. H. E. Bair for advice and
assistance in differential scanning calorimetry, and Mr. R. G.
Vadimsky for taking electron micrographs.

REFERENCES

1. R. J. Roe and C. Gieniewski, manuscript in preparation.

2. S. Matsuoka, R. J. Roe and H. F. Cole, pp. 255-271 in "Dielec-
    tric Properties of Polymers", edited by F. E. Karasz, Plenum
    Publishing Corp., N. Y. (1972).

3. R. J. Roe, H. F. Cole and E. P. Otocka, ACS Polymer Preprints
    12, 311 (1971).

4. D. J. Blundell, A. Keller and Kovacs, J. Polymer Sci., B2, 4,
    481 (1966).

5. H. E. Bair, R. Salovey and T. W. Huseby, Polymer, 8, 9 (1967);
    H. E. Bair, T. W. Huseby and R. Salovey, in "Analitical Calor-
    imetry", edited by Porter and Johnson, Plenum Press (1968).

6. H. F. Cole, PhD. Thesis, Rutgers Univeristy, 1971.

7. L. Alexander, p. 102 in "X-ray Diffraction Methods in Polymer
    Science", John Wiley & Sons (1969).

8. P. J. Flory, J. Chem. Phys. 15, 684 (1947).

9. L. Mandelkern, "Crystallization of Polymers", McGraw-Hill (1964).

10. M. J. Richardson, P. J. Flory and J. B. Jackson, Polymer 4, 221 (1963).

11. J. I. Lauritzen and J. D. Hoffman, J. Research Natl. Bur. Standards, 64a, 73 (1960); J. D. Hoffman and J. I. Lauritzen; ibid., 65a, 297 (1961); J. P. Price, J. Polymer Sci. 42, 49 (1960); F. C. Frank and M. Tosi, Proc. Roy Soc. (London) A263, 323 (1961).

12. J. D. Hoffman, J. I. Lauritzen, E. Passaglia, G. S. Ross, L. J. Frolen and J. J. Weeks, Kolloid Z. u. Z. Polymere. 231, 564 (1969).

13. J. I. Laurtizen, E. DiMarzio and J. E. Passaglia, J. Chem. Phys. 45, 4444 (1966).

14. R. J. Roe and H. E. Bair, Macromol. 3, 454 (1970).

SMALL ANGLE X-RAY DIFFRACTION FROM CRYSTALLINE POLYMERS

A. F. Burmester* and P.H. Geil

Case Western Reserve University

Cleveland, Ohio 44106

ABSTRACT

An investigation of the small angle diffraction (SAXD) patterns of various polyethylene (PE) and polyoxymethylene (POM) specimens suggests that the two discrete peaks generally produced by as-crystallized material are due to the presence of two separate diffracting structures. Electron microscopy (EM) observations of a POM fracture surface also revealed two distinct physical structures which corresponded to the two SAXD peaks obtained from that specimen. The lamellae appearing on the free surfaces of unoriented melt crystallized POM and PE correspond to the second or smaller SAXD long period (l_2). During annealing of POM near the melting temperature the structure corresponding to the l_1 SAXD peak increased its repeat distance through a "premelting" of the lamellar surfaces. Upon cooling the "premelted" material reorganized on the lamellar surfaces in a way that frequently increases the order of the repeating structure so that a second order Bragg diffraction was observed. When this occurred, the original l_2 became obscured. In some materials (nylon-11) the changes in long period may have occurred by a different mechanism. The observance of abrupt changes in the long period by an amount corresponding to the crystalline $\underline{c}$ axis repeat distance suggests that the change in long period is directly due to changes in the thickness of the crystalline portion of the lamellae.

*Present address Dow Chemical Company, Midland, Michigan.

INTRODUCTION

Small angle x-ray diffraction (SAXD) should be of considerable
value in the characterization of the lamellar texture of crystalline
polymer solids; however, the interpretation of the discrete diffraction from these materials has been the subject of controversy for
some time, both before and following the discovery of their lamellar
nature. Although, based on single crystal studies, one expects to
be able to relate the SAXD spacings to the lamellar thickness,
several observations have been made on melt crystallized material
which cast some doubt upon this interpretation and threatens to
reduce the usefulness of SAXD. These observations fall into three
separate categories: (1) those involving disagreement between the
lamellar thickness observed by electron microscopy (EM) and the
SAXD spacing or long periods calculated from the position of discrete SA scattering by Braggs Law; (1-3) (ref. 4 and 5 utilizing
nitric acid digested bulk material, ref. 6 using ion etched,
oriented samples, and ref. 7 using stained, oriented rubber suggest
agreement). (2) those involving observations of multiple discrete
SAXD peaks which could not be related to a unique spacing through
the Bragg equation: (1-3,8-10) (The necessity of properly correcting the data, as generally obtained, for slit smearing effects has
been pointed out by several authors (11,12). Using slit corrected
data they report ratios approaching 2 with values up to 2.48 being
still observed (11) however), and (3) observation of reversible
changes in position and intensity of SAXD peaks (7,9,13-22) during
annealing, which, if the classical interpretation was applied, would
require unexpectedly large, reversible changes in lamellar thickness.

EM data on lamellar thickness is usually obtained from replicas
of free or fractured surfaces. Ion etching has been used to prepare surfaces but, in our experience, usually results in localized
surface heating and annealing. One also has to consider if the
surface structure represents the interior structure, the latter
being determined by SAXD; this is particularly of concern in thick,
quenched samples. Local shadow angles on the irregular surfaces
can be determined by distributing latex particles over the surface
prior to shadowing (23). Knowledge of the angle of intersection of
the lamellae and the surface is also needed; either a parallel or
perpendicular alignment is desirable. Specially oriented samples
may be helpful. The actual measurement of step height is rather
subjective; the choice of which area, lamellae and where along a
step to measure being subject to the microscopist's experience and
bias. The optical diffraction method (8) of analyzing the microscopy results removes most of the subjective factor but can only be
used on samples in which the periodicity is parallel to the surface.

The evaluation of SAXD experiments involves 2 problems; correction of the data due to sample, camera geometry and background effects and interpretation of the corrected diffraction pattern in terms of the morphology of the sample. These two problems are considered below.

Although less desirable, intensity considerations frequently lead to use of slit collimation rather than pin hole collimation. Although the theory for corrections for the slit-smearing effect is well-known, (24,25) in actual practice the correction is rather complicated and requires computer facilities; under the apparent assumption that the relative effect (on the magnitude of ℓ_1 and the ℓ_1/ℓ_2 ratio) was small many of the results cited above were based on the directly measured diffraction scans. Recently several practically useful methods have been described for correcting for the slit-smearing effect (26-30). That used here is a modification of the method described by Schmidt and Hight (30) that is particularly suitable for computer calculation. Background corrections (including counter system noise, main beam profile, air scattering and parasitic scatter from the collimation system) suitably corrected for sample absorption are also required; this can be done most easily by subtraction of the intensity observed with the sample inserted in the beam near the entrance slit from the intensity measured with the sample in the normal position.

Two additional corrections, to our knowledge not previously applied to discrete diffraction data from bulk polymers, also are needed. The Lorentz and geometrical corrections are routinely applied in wide angle x-ray diffraction (31) and appear implicitly in the equations utilized for obtaining mass and length parameters from SAXD independent particle scattering. The geometrical factor arises from the increase in Debye-Scheerer ring diameter with diffraction angle. For small angles both corrections (32) involve multiplying the measured intensity by Θ. As shown below significant changes in the form of the diffraction scan result. For a highly oriented sample, in which the lateral extension of the reflection is less than the slit height, only the Lorentz factor would appear to be applicable.

Following correction of the data one is still faced with the problem of interpreting the results. Generally, in the past, Bragg's law has been applied, usually to the uncorrected data. However, Bragg's law is applicable only when the diffraction peaks have a width small in comparison with the diffraction angle, a situation seldom realized for melt crystallized polymers. In general one has to consider both the lattice factor and the particle factor; the paracrystallinity theory developed by Hoseman (33) would appear most appropriate. Blundell (36) and Tsvankin (37,38) have modified the theory to include a gradual change in electron density between the crystalline and interlamellar regions rather than a step function,

2 phase system.

Reinhold et al., (34) have utilized Hoseman's theory to examine the shift in ℓ_1 and ℓ_2 due to an asymmetry in the distribution of lamella thickness. One anticipates, if the lamella thickens after formation (10,35), that the average spacing would be larger than the maximum in the distribution. The application of Bragg's law to the pattern resulting from such a distribution leads to a spacing that is less than the average spacing. This is opposite the effect desired to reconcile EM and SAXD observations, the EM observed average lamellar thickness already being less than the Bragg spacing (1-3).

The above methods utilize comparison of the measured, corrected diffraction data with that calculated from a model. More direct approaches include calculation of radial distribution functions, as utilized by Blasie and Worthington (39) for biological membranes and calculation of the correlation function (40). The former still utilizes an assumed model for calculation of the scattering factor, leading to evaluation of the lattice repeat distance distribution. The latter is analogous to the Patterson function and is directly calculated from the corrected data. The physical meaning can be visualized by considering a rod of any given length placed in the scattering system with one end resting in a certain phase; the correlation function gives the probability of finding the other end of the rod in the same type of phase (i.e., same electron density), as a function of rod length.

Several other techniques can also be applied to the general problem of relating SAXD and EM measurements to the morphology of bulk crystalline polymers. These include infrared, wide angle x-ray diffraction (WAXD) and thermal measurements. Koenig et al., (51) have shown that particular absorbtion bands in the infrared spectrum of some crystalline polymers may be assigned to a regular fold conformation of the molecule. Changes in the regularity of the fold conformation, as required in premelting models for the changes in lamellar thickness, (58,61) should lead to a disappearance of these bands. WAXD can be used for crystallite size determination through line width measurements (44-46). The results can be compared with the SAXD long period to also determine the thickness of the amorphous surface layers. Likewise, since the premelting phenomen is a thermodynamic change of phase, it may be expected that its onset could be observed by thermal measurements.

The experiments reported here have been concerned with all three of the problem areas previously mentioned; the emphasis, however, has been on the reversible temperature effects in the SAXD pattern. Previously only limited information has been reported concerning this effect, particularly changes in spacing, and it was hoped that better definition of the long period behavior with

temperature would lead to a more positive interpretation of the
small angle patterns.

EXPERIMENTAL

Small Angle X-Ray Diffraction

Small angle x-ray diffraction data were obtained using a
Rigaku Denki diffractometer, RU-3H rotating anode x-ray source with
a copper target and a Hamner counting system which included a Ni
filtered scintillation counter, linear amplifier and pulse height
selector. Both continuous-scanning or step-scanning was possible.
A programmed temperature oven was constructed (47) which was capable
of maintaining sample temperatures between -100°C and 260°C. Calibration runs using indium on a mounted polymer film indicated a
difference in temperature of 0.6°C between the copper block holding
the sample and the polymer film surface. An automatic electronic
controller was used to carry out the scanning action of the diffractometer, the acquisition and recording of counting rate and scattering angle data, and the sequencing of oven temperature (48). The
data was then treated on a Univac 1108 computer as outlined below
(47).
   (1) Outlier identification and rejection, smoothing and sub-
        traction of the background scan from other designated
        scans, and smoothing of the resulting specimen scans;
   (2) Slit height correction;
   (3) Lorentz-geometrical correction;
   (4) Correlation function computation.

The slit correction method used in these measurements was a
modified version of that of Schmidt and Hight (30). Their method
involves the assumption that the vertical intensity distribution at
the collimator entrance is Gaussian with a width expressed by the
parameter, p. Instead of measuring this intensity distribution
directly, an indirect method was used. The SAXD pattern of an unoriented specimen was obtained with both pin hole and slit colli-
mations. Both patterns were corrected for background and slit
correction was applied to the slit collimated scan. Since the parameter p controls the extent of the correction the shape of the corrected pattern could be modified by varying p. The best value of
p was that which produced the closest agreement between the corrected
and the observed pin hole scan. Separate determinations of p were
made using three different scattering specimens whose SAXD patterns
varied considerably with respect to shape and position of maxima.
The values of p recovered in each case were equal to within 15%.

Both the slit correction and the geometrical correction are
designed for application to unoriented patterns; patterns from highly
oriented specimens were corrected only for background and Lorentz

effects. In principle, however, patterns obtained from partially
oriented materials would require some degree of slit and geometric
correction dependent upon the degree of orientation. Since the only
oriented specimen used in this investigation produced a rather highly
oriented pattern these corrections were neglected.

As pointed out above, the application of the Bragg equation to
broad discrete peaks in a diffraction pattern may lead to erroneous
values for the repeat distance. Therefore, in addition to the use
of the Bragg relation, correlation functions were calculated by
Fourier transforming the corrected data. Since discontinuities in
the high angle region of the diffraction pattern lead to spurious
termination errors in the correlation function, data were recorded
out to angles at which the scattered x-ray intensity was small compared to the intensity measured at the discrete peaks. In some
cases the average repeat distance of the dominant periodicity could
be taken directly as the position of the first maxima of the correlation function. On the other hand, in many cases where quantitative measurements were not required, important changes in the characteristics of the diffraction pattern could be obtained by the simple
expedient of obtaining a long period by the application of Braggs
Law to the corrected diffraction pattern. In the following discussion results reported under the title "long period" or "ℓ" were
obtained in this manner.

Electron Microscopy

All of the x-ray specimens were subjected to conventional
electron microscopy techniques. In addition several of the specimens were examined by techniques developed for the special problems
of this study. Several specimens were also prepared in a particular
manner which facilitated the microscopic study.

Surface replicas were obtained in the following manner: The
sample surface to be replicated was washed with distilled methanol
and a dilute water suspension of 860 A spherical polystyrene latex
particles was then placed on the surface and allowed to evaporate.
In some cases attempts were made to increase the wetability of the
surface, prior to placing the latex suspension on it, by a brief
exposure to a corona discharge from a Tesla coil. This resulted in
a more uniform distribution of the particles with no noticeable
effect on the microstructure of the surface. After the water evap-
orated, leaving the latex dry on the surface a small amount of hot
water was placed on the surface to dissolve the soluble soaps
present in the latex suspension. The water was removed with a micro-
pipet after a few seconds. The samples were then replicated, using
Pt-C shadowing at an angle of ca. 20°C. To eliminate magnification
errors all micrographs were obtained with the same lens settings; a
replica of 54,800 line per inch ruled grating was also photographed

each time. Later, during photographic enlargement of the micrographs, enlargements of the grating images were also made under the
same conditions to provide for calibration of the final, enlarged
micrograph.

Free surface replicas were obtained from nearly all of the
specimens used in the study. Since the surface structure may not
be representative of the bulk material several different methods of
attempting to observe the interior structure of the specimen were
made. Attempts to use ion etching were unsuccessful. Fracture
surfaces were prepared by immersing the sample in liquid nitrogen
for a few minutes and then fracturing it by twisting it between
two pairs of pliers. This method produced mixed results. With some
specimens it produced excellent replicas; in other cases excessive
deformation had taken place during the fracture so that no information concerning the original structure could be obtained. The best
samples are obtained from slowly crystallized samples that have been
adequately cooled. Shish-kebob type samples, in which the lamellae
are ca. perpendicular to the surface, were prepared by drawing ice
water quenched, 1/8" thick compression molded films at 130°C and then
annealing them while held taut in either air or glycerine. Samples
prepared in this manner, although showing large areas in which the
lamellae were oriented nearly perpendicular to the exposed surface,
produced a very irregular surface which somewhat hampered the shadow
method of thickness measurement.

Spit (49) and Keller and Machin (50) have demonstrated the usefulness of gold decoration for polymers, the ca. 25 A gold particles
tending to collect on the lamellar edges while the fold surface remains relatively devoid of gold. When the lamellae are on edge,
two rows of gold particles form (50). After the gold had been evap-
orated onto the surface, it was carbon coated and gold particles
removed by stripping with polyacrylic acid. One practical advantage
was that the gold decoration could be stripped from the surface with
relative ease; the Pt-C replicas adhered to the surface so tightly
that they often pulled fibers from the surface or, in some cases,
could not be removed. This was especially true of the specimens
which had been subjected to ion etching.

One of the main goals of this research was to identify
the cause of the reversible changes in the small angle pattern with
temperature changes. Thus an attempt was made to observe directly
the morphological changes which were occurring reversibly, by
forming a Pt-C replica while the specimen was held at an elevated
temperature. The sample was placed on a temperature controlled
block, using silicon vacuum grease for thermal contact, in the evap-
orator. Removal of the replica from the surface was done at room
temperature.

Studies of the irreversible effects of annealing on the morphology were conducted by EM observation of the same morphological
structure before and after the thermal treatment using a repeat
replica technique (51). Care was taken to insure that the angle
and direction of shadowing were the same as for the first replication.

In one series of experiments the last two techniques described
were combined. After annealing and lifting of the second replica,
the specimen was placed on the heated aluminum block in the vacuum
chamber and a third replica deposited at the anneal temperature.

Measurements of the infrared fold bands, as a function of
specimen temperature, were made with a Perkin-Elmer model 521 infrared spectrometer. The specimen, in the form of a thin film, was
placed in the same oven used for the SAXD studies. Mylar windows
were replaced with 1/8" thick KBr windows for the high temperature
measurements when evacuation of the chamber was required. Measurements with specimen temperature as low as −75°C were obtained by
the cold nitrogen gas technique with the windows removed to prevent
water condensation. The specimen chamber was placed in the spectrometer in such a way that the sample beam passed through the assemply
in the same manner that the x-ray beam did in the small angle experiment. Unfortunately the geometry of the specimen chamber did not
allow placement of the specimen at the normal position for the
spectrometer. As a consequence only about one half of the beam
passed through the specimen thus reducing the instrument sensitivity.
Since the features of the infrared spectrum became less distinct at
high temperatures, this loss of sensitivity became an important
limitation of the experiment. In order to make maximum use of the
available sensitivity the spectra were recorded at the slowest available scan speed and with a 5 times scale expansion along the absorbance axis. Further gain in sensitivity at the expense of some
resolution, was obtained by using spectrometer slits which were
opened to twice the normal widths.

Wide angle x-ray diffraction data were obtained at temperatures
between 25°C and the melting point of the specimen with a General
Electric XRD-6 diffractometer equipped with a heated sample cell,
with reflection geometry (47).

Measurements of the thermal properties of some specimens were
obtained with a DuPont Model 900 Differential Thermal Analyzer. Melt
crystallized material was ground in a ball mill for one minute after
it had been chilled in liquid nitrogen. Thermograms obtained from
samples prepared in this manner were the same as those obtained from
samples obtained by shearing the bulk material with a knife. The
samples were annealed for one hour at verious temperatures in the
DTA, cooled to room temperature and then heated through the melting
point at a rate of 25 degrees/min. while the thermogram was recorded.

RESULTS AND DISCUSSION

The data described below is divided into two categories: (1) SAXD and EM measurements taken on the same specimen, and (2) annealing effects. These two categories correspond closely with two of the three SAXD interpretation problem areas. The remaining area, that of the relationship between multiple peaks in the SAXD pattern, is of a general nature and data pertinent to it are presented in both parts of this section.

EM and SAXD Measurements

In the initial stages of the research a number of polymers were examined in order to select a material which produced strong SAXD maxima and also possessed morphological features which could be replicated and measured for EM. These materials were also subjected to SAXD measurements at elevated temperatures in order to establish if the reversible changes in SAXD pattern previously reported for some materials (15,16) were general phenomena. Surprising was the number of materials which were found to be unsuitable. For example, although polyethylene sulfide, 1-4 polyisoprene and poly-4-methyl pentene-1 all produced excellent free surface replicas showing extensive lamellar structure, their SAXD pattern consisted of weak shoulders on the diffuse scatter or no discrete peak at all. At elevated temperatures, the intensity of all increased, but insuf-ficiently to permit their use. On the other hand the SAXD pattern of polypropylene showed a rather well defined maxima but the surface morphology was an irregular array of narrow, ribbon-like structures. Materials selected for further study were prepared and identified as shown in Table 1. LPE and POM were used in both the morphological and the annealing experiments. PET and nylon 11 were used only in the annealing studies.

A micrograph of the free surface of LPE is shown in Figure 1. The shadow angle was determined through measurement of the latex particle and its shadow to be $\tan^{-1}(2.7)$. Lamellar step heights were calculated by measurement of the shadow lengths at points in-dicated by the arrows in the micrograph, parallel lines being drawn along the visually estimated edges of the shadows. Several factors were considered in the selection of the areas to be used for analysis. A section of the spherulite in which the lamella edges were spaced relatively far apart was selected to insure that individual lamellae were oriented nearly parallel to the surface, thus eliminating one of the variables of the problem. Measurements were limited to the immediate vicinity of latex particles casting useable shadows. The shadow angle determined by measurements on the particle then accur-ately represented the angle at the site of the lamellar edge. Even

TABLE 1

SAMPLE IDENTIFICATION

Sample Designation	Material	Crystallization Conditions
LPE-Q	Polyethylene	Cooled from melt at 200°C/min.
POM-Q1-4	Polyoxymethylene	Cooled from melt at 200°C/min.
POM-S-1,2	Polyoxymethylene	Cooled from melt at 3°C/min.
POM-D	Polyoxymethylene	Same as POM-Q then drawn at .05 in/min at 130°C and annealed with ends fixed at 163°C for 1 min.
POM-I	Polyoxymethylene	Crystallized isothermally from the melt at 3°C/min.
N-11-S	Nylon-11	Cooled from melt at 3°C/min.
PET-A	Poly(ethylene terephthalate)*	1 mil. amorphous film crystallized by annealing at 180°C/1 hr.

*Furnished through the courtesy of
 C. Heffelfinger, E. I. DuPont Company.

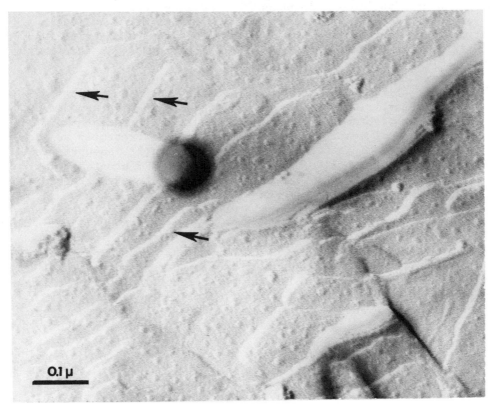

Figure 1   Electron micrograph of a free surface of LPE-Q showing
            latex particle used for determining local shadow angle.

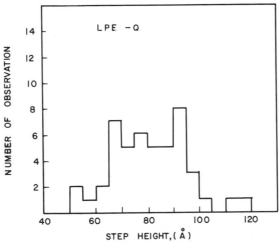

Figure 2   Histogram of lamellar step heights measured from micro-
            graphs such as Figure 1 of sample LPE-Q.

though an area was chosen in which the lamellae were well spaced, two or more shadow lines were frequently seen to merge indicating that the step represented more than one lamella at that point. An example of this is seen just below the latex particle in this micrograph.

The probable error in the measurement of both the shadow angle at the latex particle and of individual shadows was between 5 and 10% depending on the shadow definition and length. However, the deviation of individual measurements from the mean was freqently as much as 50%.   Figure 2 is a histogram of the step height measurements made on LPE-Q for a total of 47 measurements on different shadows. The distribution is roughly centered about 80 A; however, the uniformity of the distribution is such that the results are presented in Table 2 in terms of a range of lengths. In this particular case the lamellar step height was found to be between 65 A and 95 A.

The SAXD pattern of LPE-Q with the various corrections applied is shown in Figure 3. The uncorrected data corresponds to that obtained using a continuous diffractometer scan. A poorly resolved discrete peak with only a bare suggestion of a second peak is present. Arrows placed over the curve indicate the peak position determined by drawing in an estimated diffuse scatter background. The subtraction of an actual background scan improved the definition of the diffraction peak considerably; however, it remained a shoulder with no minima on the low angle side. Correction for slit smearing effects altered the shape and position of the ℓ_1 (lower angle) peak considerably. The definition of the ℓ_2 peak was also improved although its position did not change appreciably. Finally, application of the Lorentz-geometrical correction (multiplying by θ^2) further changed the shape of the entire curve. Not only was the definition of both of the observed peaks improved but the positions of both were shifted to higher angles (ℓ_1 more than ℓ_2).

The correlation function calculated from the fully corrected SAXD pattern for LPE-Q is given in Figure 4. Since the correlation function is an indication of the probability of finding volume elements of equal electron density separated by a given distance, maxima in the function may be interpreted as an indication of characteristic physical repeat distances within the specimen. The fact that this function is the Fourier transform of the diffraction pattern means that each feature of the function (e.g., the first neighbor peaks) contain information from the entire diffraction pattern. Hence an average repeat distance based on the location of a peak, such as appears in Figure 4 at 247 A for LPE-Q, is probably more accurate than the long period obtained through the application of the Bragg equation to isolated features of the diffraction pattern. The shape of the correlation function peak can also be interpreted in terms of the distribution of repeat distances. In this case the first neighbor peak was rather broad and somewhat asymmetrical, and

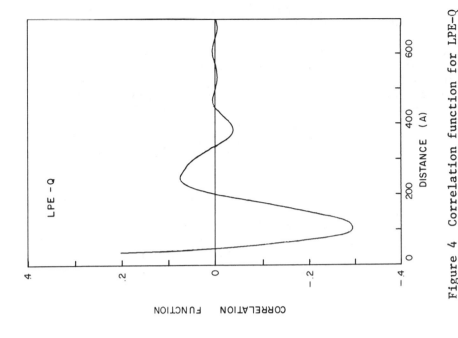

Figure 4   Correlation function for LPE-Q calculated from the corrected pattern shown in Figure 3.

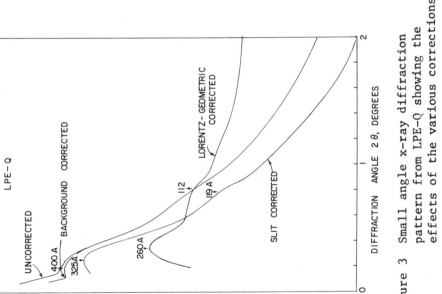

Figure 3   Small angle x-ray diffraction pattern from LPE-Q showing the effects of the various corrections.

no significant second neighbor peak appeared. This is equivalent
to the observation that only a poor "second order" Bragg reflection
appeared in the diffraction pattern.

One important disadvantage of measurements of surface features
(such as above) is that these features may not be representative of
the features within the bulk of the material. In order to obtain
information concerning the interior structures, attempts were made
to observe features on fracture surfaces. Only POM-I yielded satisfactory fracture surfaces. Two distinct types of surfaces were
observed. One of these (Figure 5a) consisted of spherical particles
arranged in more or less distinct rows on a relatively flat surface.
These are areas in which the lamellae were presumed to be oriented
nearly perpendicular to the fracture surface with the row-like
arrangement of the particles indicating the direction of the underlying lamellae. It has been suggested that these particles result
from local melting and retraction of deformed material during the
fracture process (1,52). The size of the particles varied considerably but there seemed to be a characteristic lower limit. Determinations of lamellar thickness were made from this type of surface
under the assumption that short rows of the smallest of the particles
represented individual underlying lamellae. The other type of structure observed in this fracture surface replica is shown in Figure 5b.
It resembles regions of a free surface in which the lamellae are
roughly parallel to the surface. Distinct lamellar edges were observed and the methods described in connection with Figure 3 were
applied to determine lamella thicknesses. The values of lamellar
thickness determined from this surface were considerably smaller
than those measured from the surface where the lamellae were oriented
perpendicular to the surface. One possible reason for this is that
even the smallest of the particles seen in Figure 5a do not represent single lamellae but instead represent material pulled out of
two or more adjacent lamellae at the time of fracture. However, in
a study of poly-4-methyl pentene-1 in our laboratory (53) two distinct lamellar structures on a fracture surface have also been observed. The lamellae which were oriented perpendicular to the surface were clearly defined by boundary lines and were approximately
350-400 A thick while those lamellae oriented parallel to the surface had a step height of about 150 A.

The SAXD pattern for this POM-I with the various corrections is
given in Figure 6. This pattern is unusual in that the corrected
intensity of the second (ℓ_2) peak is higher than the ℓ_1 peak. A
weak and quite broad peak is also present at a scattering angle of
about 1 degree corresponding to a Bragg spacing of 90 A. The
correlation function for POM-I contains two peaks; one at 170 A and
the other at 312 A (Figure 7). Since the second is not at a distance
of twice the first and also is higher than the first, it could not
be considered as a second neighbor peak. Instead the second peak,
we suggest, indicates the presence of a second, distinct, physical

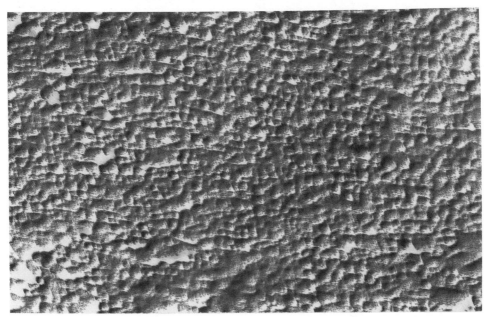

Figure 5a   Electron micrograph of a fracture surface of POM-I.
Lamellar spacing is 220–320 A.

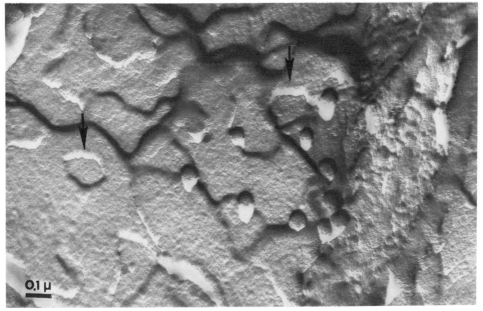

Figure 5b   A different region of the same surface as shown in
Figure 5a.   Step height, as at arrows, is 140–170 A.

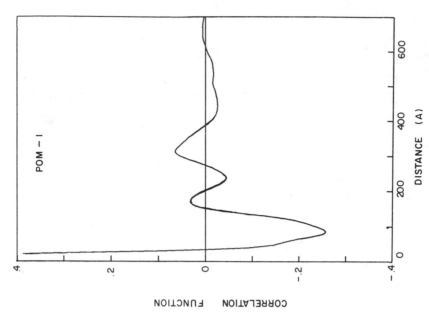

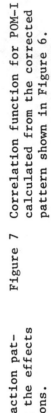

Figure 7   Correlation function for POM-I calculated from the corrected pattern shown in Figure 6.

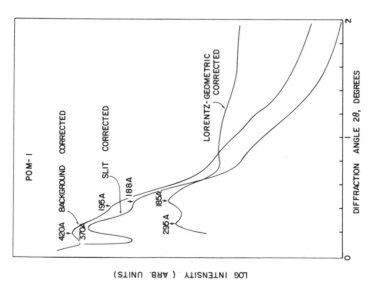

Figure 6   Small angle x-ray diffraction pattern from POM-I showing the effects of the various corrections.

repeat distance within the scattering specimen.

    As previously indicated, one of the possible sources of the
discrete SAXD is an extended array of stacked lamellae. Such
lamellar arrangements are not normally observed on the free surface
of melt crystallized polymer but can be observed on drawn, annealed
samples (54). Specimen POM-D was prepared in this manner so that
measurements of the lamellar periodicity could be made directly by
viewing them edge on. In such regions the measurement includes not
only the crystalline lamellar thickness but also the thickness of
any interlamellar layer. Figure 8a is a micrograph of the surface
of POM-D replicated using C-Pt shadowing at an angle of about 45
degrees. Areas in which the lamellae are presumed to be oriented
perpendicular to the surface are poorly defined by this shadowing
technique. A micrograph of a similar area of the same specimen
that had been gold decorated is shown in Figure 8b. In many
locations, indicated by arrows, the rows of Au particles are seen
to be closely parallel over extended distances. These are assumed
to be regions in which several lamellae were oriented approximately
normal to the surface. The repeat period of the lamellar stack
listed in Table 2 was determined by measuring the spacing of the
lines of particles.

    The SAXD data obtained from POM-D are plotted in Figure 9. In
this case two well defined, discrete maxima were observed, even in
the uncorrected scan. The scattering from the specimen was much
stronger than the background in the area of the first peak and consequently, the background connection produced very little change in
the character of the diffraction pattern. Only the Lorentz correction
was applied; since the peaks were rather sharp this correction
shifted both peaks only slightly. The correlation function calculated from the SAXD pattern for POM-D appears in Figure 10. Here
the two maxima in the function were in the proper relation regarding
position and relative height to be interpreted as first and second
neighbor peaks from a single physical repeat structure. The very
deep first minima and the appearance of two related maxima indicate
a better degree of order or regularity of the repeat distance than
observed in the previously discussed specimens.

    The numerical results of these EM and SAXD measurements as well
as similar measurements on rapidly cooled and slow cooled polyoxymethylene specimens (POM-Q1 and POM-S1), are presented in Table 2.
In this table, the SAXD data are presented as the angle of the discrete peak in degrees followed by a slash and the long period in
angstroms calculated by applying the Bragg equation to the angle of
the peak. The ratio of the first (lowest angle) long period to the
second (ℓ_1/ℓ_2) is also given for all of the specimens under the
various corrections. Since the corrections always shifted the first
peak more than the second this ratio decreased to lower values as
the corrections were applied. The figures listed under "Correlation

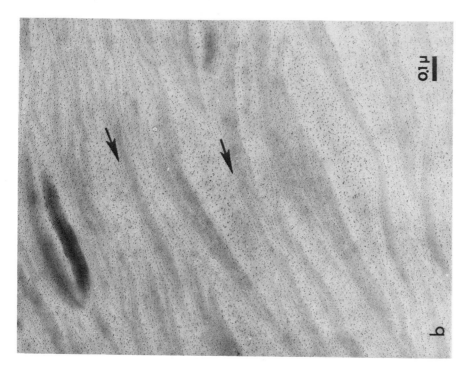

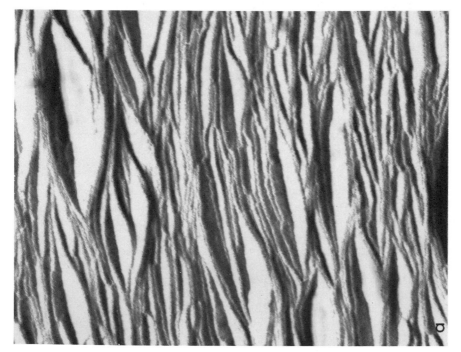

Figure 8   Electron micrograph of the surface of POM-D (draw direction vertical) with: a) Pt-C shadowing and, b) Gold decoration.

TABLE 2

SUMMARY OF SAXD AND EM MEASUREMENTS AT ROOM TEMPERATURE

Specimen		Corrections		Lorentz Geometric	Correlation Function Peaks (Å)	Electron Microscopy (Å) Measurements
		Background	Slit			
LPE-Q	ℓ_1	0.22/400	0.27/325	0.34/260	245	
	ℓ_2	0.76/116	0.76/116	0.79/112		65 to 95
	ℓ_1/ℓ_2	3.5	2.8	2.3		
POM-Q1	ℓ_1	0.43/205	0.49/180	0.62/142		
	ℓ_2	1.54/57	1.56/56	1.58/56		30 to 50
	ℓ_1/ℓ_2	3.6	3.2	2.5		
POM-I	ℓ_1	0.21/420	0.24/370	0.30/295	170,310	220 to 320 (lamellae perpendicular to surface)
	ℓ_2	0.45/195	0.47/188	0.48/185		140 to 170
	ℓ_1/ℓ_2	2.2	2.0	1.6		(lamellae perpendicular to surface)
POM-D	ℓ_1	0.53/165		0.56/155	150,325	100 to 130
	ℓ_2	1.43/62		1.55/56		
	ℓ_1/ℓ_2	2.7		2.8		
POM-S1	ℓ_1	0.37/235	0.42/210	0.55/160		
	ℓ_2	1.42/62	1.43/62	1.50/59		55 to 85
	ℓ_1/ℓ_2	3.8	3.4	2.7		

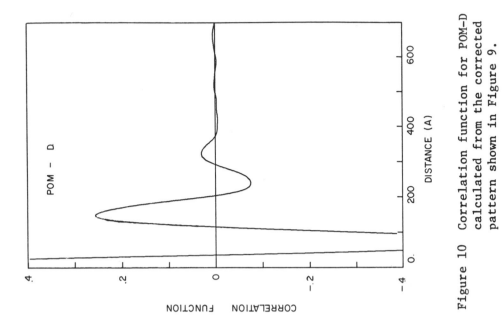

Figure 10   Correlation function for POM–D calculated from the corrected pattern shown in Figure 9.

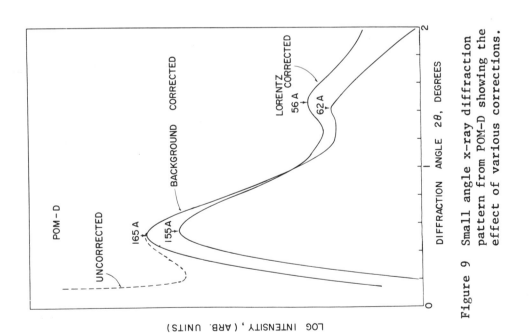

Figure 9   Small angle x-ray diffraction pattern from POM–D showing the effect of various corrections.

Function" represent the spacing of the maxima observed in the correlation function which had been calculated from the fully corrected
SAXD pattern. The two figures given for the electron microscopy
results represent the thickness range in which significant numbers
of lamellae were observed. In the case of the fractured polyoxymethylene specimen (POM-I) the two ranges represent the two different
types of surface structure found in Figure 5.

     In view of the large effects caused by the various corrections
to the SAXD patterns, it was concluded that the application of these
corrections was an important step in the treatment of the data.
Accordingly all of the SAXD patterns and results based on such patterns presented subsequently in this paper have been corrected.

Effects of Specimen Temperature on SAXD

     The SAXD patterns of several specimens were obtained at different
specimen temperatures. Figures 11a and 11b show the patterns obtained from a sample of slowly crystallized polyoxymethylene POM-S2.
In this and subsequent figures in which SAXD patterns are presented
directly, the symbols are placed on the solid curves for indentifi-
cation and do not represent individual data points. During this
experiment the recording of the diffraction pattern required a
period of sixty to ninety minutes. After the pattern had been recorded the temperature was changed to the next value and a period of
sixty minutes was allowed for thermal equilibrium and annealing
processes before the next scan was begun. Normally the experiment
consisted of twenty separate scans including a background. The ten
patterns shown here are representative of the complete experiment.
Data from all of the scans were used for the summary plots to be
presented later.

     The original room temperature pattern consisted of one well
defined peak at 0.52 degrees (ℓ_1) and one very broad and weak peak
centered at about 1.30 degrees (ℓ_2). The patterns obtained at increasing temperatures indicate that initially the intensity of the
ℓ_1 peak increased by a factor of about 3 while the intensity of the
ℓ_2 maxima remained at nearly the same level. Note that this differs
from the results of Aoki et al.,(19) for an annealed sample; in their
case both intensities increased equally. The increase seen at 124°C
is believed due to a rise in the background scattering. Above 140°C
the ℓ_1 peak became sharper and began moving to smaller angles and
at 150°C, which is 13°C below the melting temperature, the intensity
of this peak reached a maximum, decreasing slightly with increasing
temperature above this point as well as continuing to move to
smaller angles. The ℓ_2 peak remained essentially unchanged, even
at specimen temperatures only 2°C below the melting temperature. As
the sample was cooled in steps, as for the heating, the ℓ_1 peak
shifted to larger angles, coming back to about the original angle

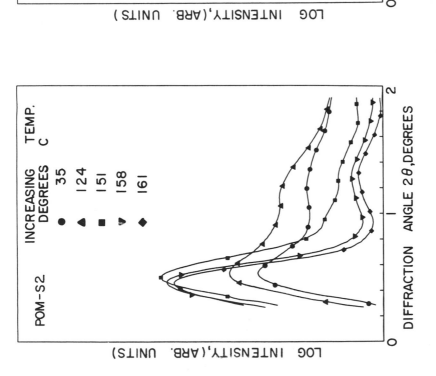

Figure 11 Small angle x-ray diffraction patterns from POM-S2 at various temperatures.

at the lowest temperature. The intensity was again observed to go
through a maximum at about 150°C. However, although the spacing
change was almost reversible, the ℓ_1 peak was considerably sharper
after annealing than it was before annealing.

The ℓ_2 peak, while showing almost no change during heating, did
change significantly during the cooling phase of the experiment.
Scans taken as the sample cooled from 161°C to 155°C showed virtually
no change in this peak. However, in the scan at 151°C the peak had
decreased in intensity and broadened considerably to the extent that
it could no longer be considered an observable feature. The next
scan (at 124°C) revealed a strong, well defined peak at 0.96° which
shifted to wider angles and generally decreased in intensity as the
specimen temperature decreased.

The correlation function computed from four of the curves in
Figure 11 are plotted in Figure 12. They show reversible shift in
the position of the initial peak and the development of a second
neighbor peak at about 300 A. The source of the relatively constant peak at 500 A is unknown. The small bump in the first minima
of the 35°C curve is due to the ℓ_2 maxima at a Bragg spacing of 70A.

Figures 13 to 15 summarize the results of this experiment
followed by one in which the same specimen was heated in ten steps
to 162°C with scans taken at each temperature and then cooled rapidly
to room temperature with a final pattern recorded at that point.
The long period (ℓ_1) using Bragg's law is plotted as a function of
specimen temperature in Figure 13. The value of ℓ_1 is seen to increase continuously from 164 A at room temperature to 196 A at a
temperature 2 degrees below the melting point. The ratio of the
intensity of the pattern at the position of the ℓ_1 peak to the
corresponding intensity of the original pattern (relative intensity)
is plotted as a function of temperature in Figure 14. The ratio
is seen to reversibly increase continuously from room temperature
to about 10°C below the melting temperature and then decrease
slightly. Figure 15 shows the initial temperature independence of
ℓ_2 during the first heating and its discontinuous change on cooling.

Data indicated by squares on these three summarizing figures
resulted from reheating the same specimen to 162°C. Within the
limits of experimental error most of the points on each plot fall
on a single line except for the initial curve for ℓ_2. It should be
noted that the ℓ_2 reheating data extend this line to temperatures
higher than the point where the shift took place during the first
cooling. Data from the pattern obtained after the specimen had
been cooled rapidly from 162°C (represented by a cross) agree with
the values obtained on the specimen after a gradual cooling from
161°C.

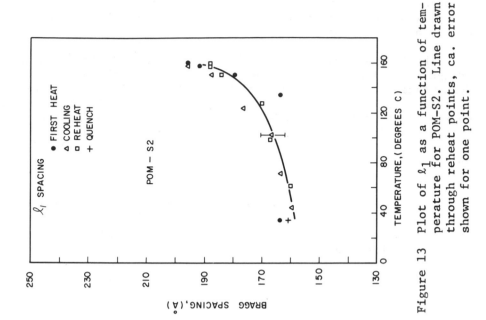

Figure 13   Plot of ℓ_1 as a function of temperature for POM-S2. Line drawn through reheat points, ca. error shown for one point.

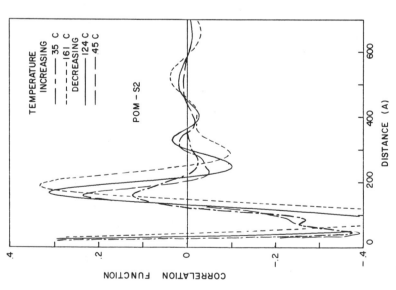

Figure 12   Correlation function for POM-S2 at various temperatures calculated from some of the patterns shown in Figure 11.

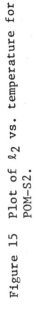
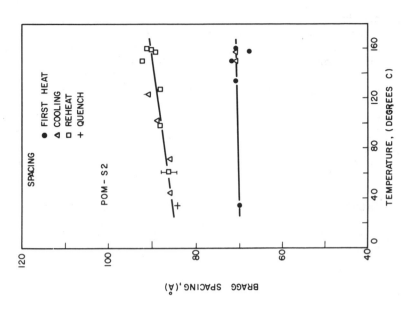

Figure 15   Plot of l_2 vs. temperature for POM-S2.

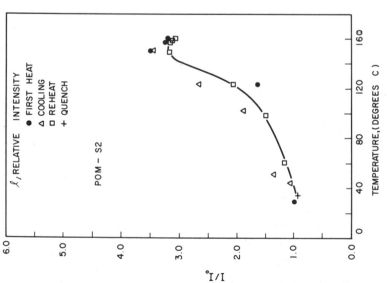

Figure 14   Plot of l_1 relative intensity vs. temperature for POM-S2. Line drawn through reheat points.

Figures 16 through 19 present data resulting from an experiment similar to the one described above for a single heating-cooling cycle that was conducted on a polyoxymethylene specimen which had been crystallized by rapid cooling (POM-Q2). The results summarizing the behavior of the ℓ_1 peak intensity and the ℓ_2 period are almost identical to those obtained in the previous experiment. On the other hand the behavior of the ℓ_1 period with changing specimen temperature (Figure 17) differs considerably from the POM-S2 case. As would be expected, the initial value of ℓ_1 was considerably less than that for POM-S2 (142 A vs. 164 A) and, although the slope of ℓ_1 vs. temperature curves was about the same up to 130 °C, the increase in ℓ_1 between 130°C and 150°C was much larger. The remainder of the heating and all of the cooling data points can, with a vertical shift of -10 A, be superimposed on the ℓ_1 vs. temperature curves for POM-S2.

The temperature cycle experiment was also carried out on an oriented polyoxymethylene specimen which had been annealed for one minute near the melting point (POM-D). This initial annealing resulted in the formation of the lamellae shown in Figure 8. The representative SAXD scans presented in Figure 20 at first sight show some similarities to the annealed, unoriented specimen (POM-2, Figure 16) in general shape and behavior with specimen temperature. There were however, some significant differences: (1) the definition of the ℓ_1 and ℓ_2 peaks in the original low temperature scan was much better than the peaks of the previous specimens. (2) The behavior of the original ℓ_2 peak was also different in that the peak began to deteriorate at 100°C and was nearly unresolvable at 147°C. With further increase in specimen temperature ℓ_2 again appeared, well resolved, at a slightly smaller angle than before. Cooling the material to 155°C from 162°C caused the same abrupt shift of the ℓ_2 peak to a smaller angle as seen for POM-S and Q. Further cooling caused a pronounced outward shift of this peak as well as the formation of a poorly defined third peak. After the return to lower temperatures the shape of the ℓ_1 peak was considerably distorted by a shoulder on the low angle side of the peak.

The summary plots for POM-D, shown in Figures 21-23, illustrate the considerable differences between these and the previously described results. No increase is seen in ℓ_1 until the specimen temperature exceeded 140°C. Between 140°C and 160°C, ℓ_1 increased from 155 A to 193 A, a change of 28%, which is more than the ℓ_1 peak of POM-S2 shifted over the entire temperature range. As in the previous cases, the maximum intensity of the ℓ_1 peak changed continuously with changing temperatures. The relative intensity increase from 50°C to 161°C was 500% which is also considerably more than the 300% increase observed for the unoriented specimen. The ℓ_2 spacing (Figure 23) characteristics were essentially the same as seen previously.

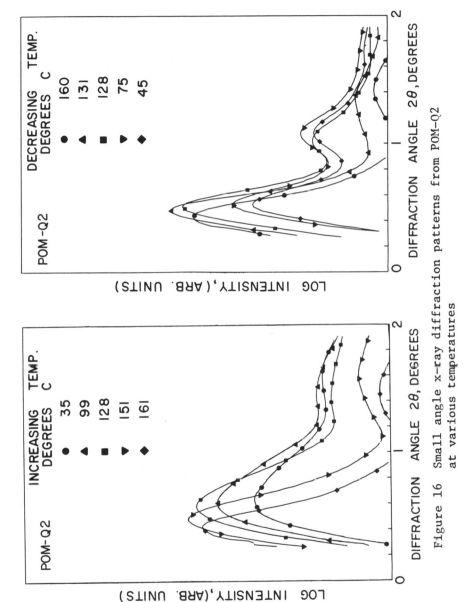

Figure 16   Small angle x-ray diffraction patterns from POM-Q2
at various temperatures

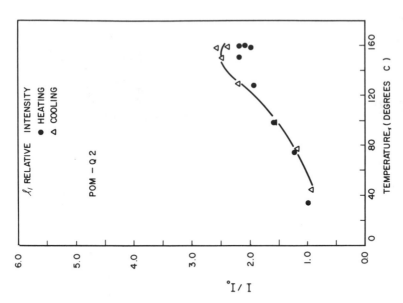

Figure 18   Plot of l_1 relative intensity as a function of temperature for POM-Q2. Line drawn through cooling points.

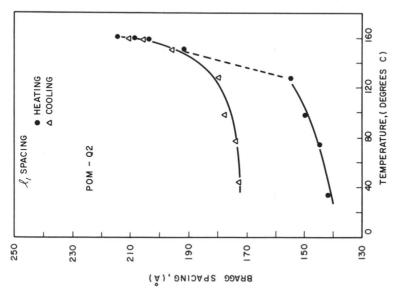

Figure 17   Plot of l_1 as a function of temperature for POM-Q2.

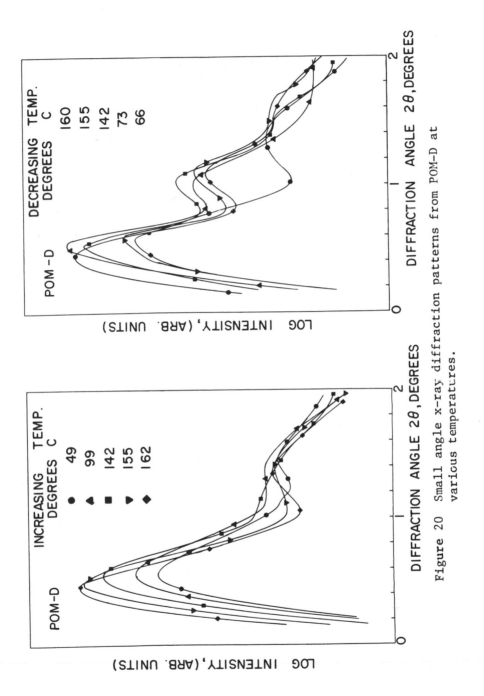

Figure 20   Small angle x-ray diffraction patterns from POM-D at various temperatures.

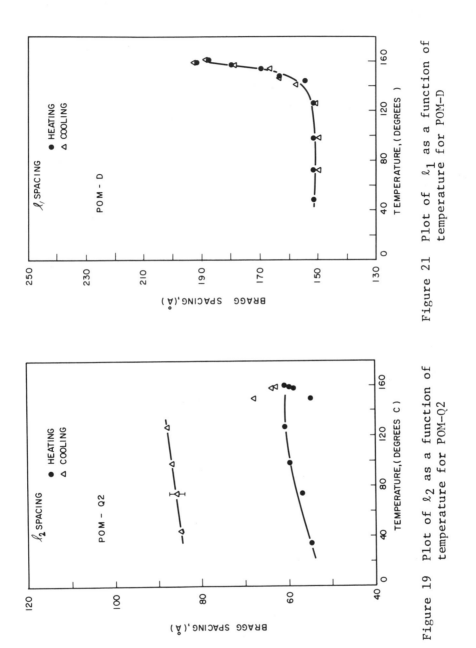

Figure 21   Plot of ℓ_1 as a function of temperature for POM-D

Figure 19   Plot of ℓ_2 as a function of temperature for POM-Q2

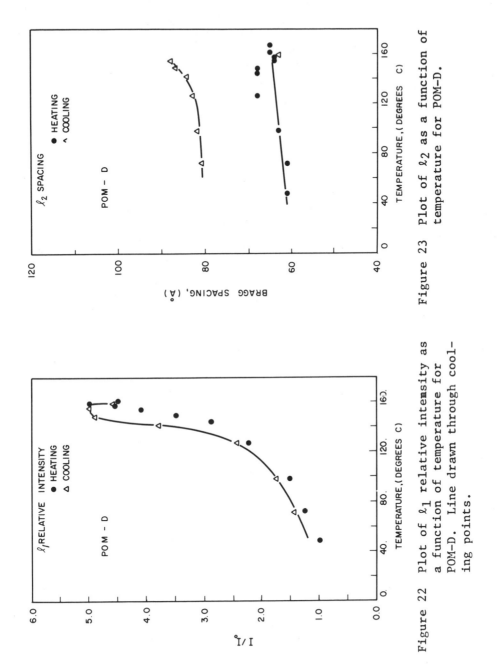

Figure 23 Plot of ℓ_2 as a function of temperature for POM-D.

Figure 22 Plot of ℓ_1 relative intensity as a function of temperature for POM-D. Line drawn through cooling points.

To examine the relationship between ℓ_1 and ℓ_2 the ratio ℓ_1/ℓ_2 was plotted as a function of temperature for the three experiments presented above. This plot, shown in Figure 24 shows that in all three cases the initial value of the ratio was between 2.3 and 2.6. Upon heating two factors caused shifts in the ratio. Between 120°C and 150°C a loss in the resolution of ℓ_2 caused some uncertainty in the determination of its value (see Figure 15 and 19). Above 140°C the dramatic increase in ℓ_1 was not matched by a shift in ℓ_2. Consequently the value of the ratio increased. After the step change in ℓ_2 at about 145°C on first cooling, the value of the ratio was 1.85 to 2.05 and remained constant even when the temperature of the specimen was again increased to near the crystalline melting point.

SAXD patterns were obtained as a function of specimen temperatures from a number of different materials. Figures 25 and 26 summarize the data obtained using a specimen of slowly crystallized nylon-11 (N-11). The pattern consisted of a poorly defined shoulder on a broad diffuse scattering component. The entire pattern was very much weaker than that observed for polyoxymethylene, making observations of second maxima unlikely even if they existed. Increased specimen temperature caused a significant decrease in the diffuse scattering, allowing better resolution of the discrete peak. In contrast to the behavior of the POM ℓ_1 peak, the total intensity at the position of the N-11 ℓ_1 peak decreased irreversibly during the course of the experiment. This decrease, which occurred at specimen temperatures above 100°C, may be attributable to the considerable reduction of the diffuse scattering background. The behavior of the long period also showed some significant differences from the data of POM-2. While the total relative change from room temperature to just below the melting point was nearly the same, the shape of the ℓ vs. T curve was different in two obvious aspects: (1) The increase in the N-11 long period was not continuous but rather occurred in at least two distinct steps. (2) At the highest temperature, the long period remained constant over an eight degree range whereas the long periods of the other polymers examined increased sharply at temperatures just below the melting temperature. The decrease of the long period during decreasing temperature also proceeded with a distinct step at 160°C followed by a gradual further decrease. A repeat of the temperature cycle produced data which fell, during both heating and cooling, onto the decreasing temperature curve shown here. Discontinuous increases in the long period in single crystals of several nylons (55) and of relatively low molecular weight polyethylene oxide (56) have been described.

In order to determine the effect of time at the annealing temperature on the changes in the SAXD patterns described, an experiment was conducted in which POM samples were annealed at the same temperature for varying periods. POM-Q4 was divided into four samples. One of these was placed into the x-ray sample chamber and the initial room temperature pattern recorded. For the individual

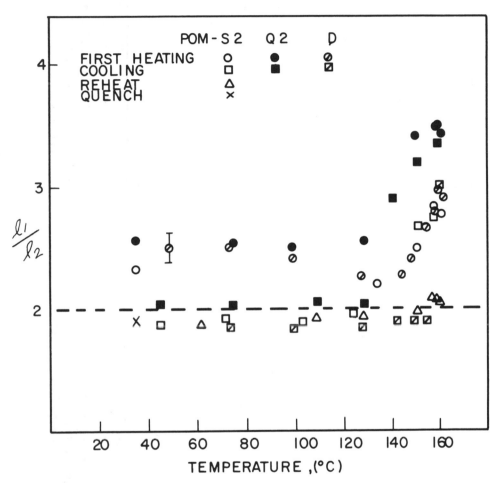

Figure 24   Plot of ℓ_1/ℓ_2 as a function of temperature for several samples.

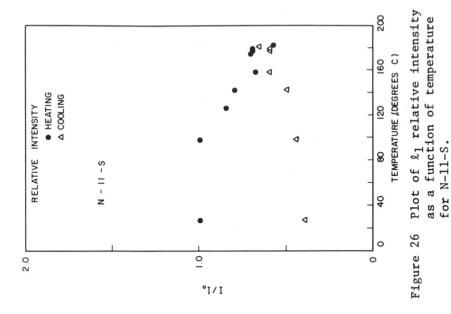

Figure 26   Plot of ℓ_1 relative intensity as a function of temperature for N-11-S.

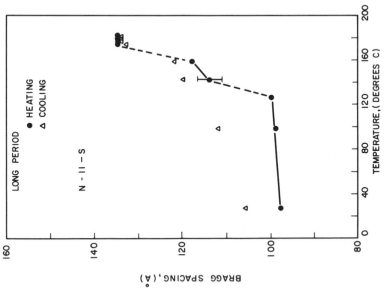

Figure 25   Plot of ℓ_1 as a function of temperature for N-11-S. Line drawn through heating points.

samples the temperature was then raised to 159°C for a period of
from 2 minutes to 16 hours. After the sample had been annealed for
a specified period the SAXD pattern was recorded, the temperature
quickly reduced to about 25°C, and a third diffraction pattern was
obtained at room temperature. An example of the data obtained in
this experiment appears in Figure 27. The total annealing time in
this case was seven hours. Both reversible and irreversible effects
on the position and shape of the two discrete peaks are apparent.

The SAXD patterns obtained at the annealing temperature after
various periods are shown in Figure 28. A large shift, considerable
sharpening, and a two fold increase in intensity occurred in l_1.
These changes required less than 5 minutes. Patterns at the annealing temperature representing annealing periods of less than 5 minutes
were not obtainable since the minimum time required to obtain a scan
of the l_1 section of the pattern is 10 minutes. Patterns obtained
on samples annealed for longer periods show no additional change in
l_1 peak position and only a minor amount of peak sharpening and
intensity gain. The l_2 peak, which was initially unobserved, was
observed in the first scan over the entire range of angles (the
actual time of anneal for the l_2 region of the pattern was approximately 60 minutes since the entire scan required one hour to complete). With increased annealing time the l_2 peak moved to slightly
smaller angles without appreciably changing shape or intensity.
Figure 29 presents the patterns obtained from the same samples as
shown in Figure 28 after they had been cooled to room temperature
plus a representative preanneal scan. Again it can be seen that the
major changes in the l_1 peak occur with short anneal times and that
changes in the l_2 peak continue to occur after annealing times of
seven hours. The reversible portion of the changes in the l_1 peak
was essentially independent of annealing time while the irreversible
changes required an annealing time of seven hours to be completed.
No real reversible changes were observed in the original l_2 peak
and irreversible shifts were still occurring at anneal times of up
to 18 hours.

The corrected l_1 and l_2 values from both the during and after
annealing patterns are tabulated as a function of annealing time in
Table 3.

In order to correlate annealing changes in the SAXD pattern
with changes in the morphology such as those reported by Seigmann
and Geil (51), a particular area of the surface of slowly crystallized polyoxymethylene specimen was replicated both before and after
annealing. SAXD patterns were obtained before heating, after 5 hours
at the annealing temperature and at room temperature after annealing.
Annealing temperatures ranged from 130°C to 161°C. Figures 30a and b
show the same area of POM-7 before and after annealing at 161°C
respectively. The SAXD changes are seen to have occurred during the
course of the annealing treatment. Examination of the micrographs

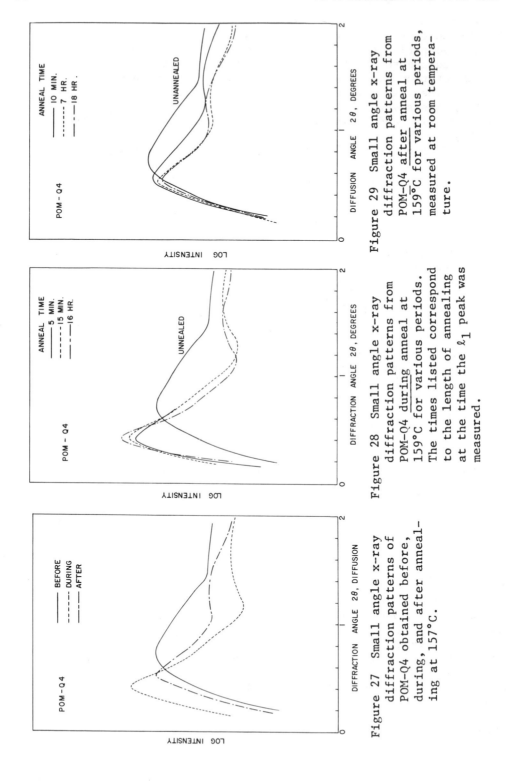

Figure 29 Small angle x-ray diffraction patterns from POM-Q4 after anneal at 159°C for various periods, measured at room temperature.

Figure 28 Small angle x-ray diffraction patterns from POM-Q4 during anneal at 159°C for various periods. The times listed correspond to the length of annealing at the time the ℓ_1 peak was measured.

Figure 27 Small angle x-ray diffraction patterns of POM-Q4 obtained before, during, and after annealing at 157°C.

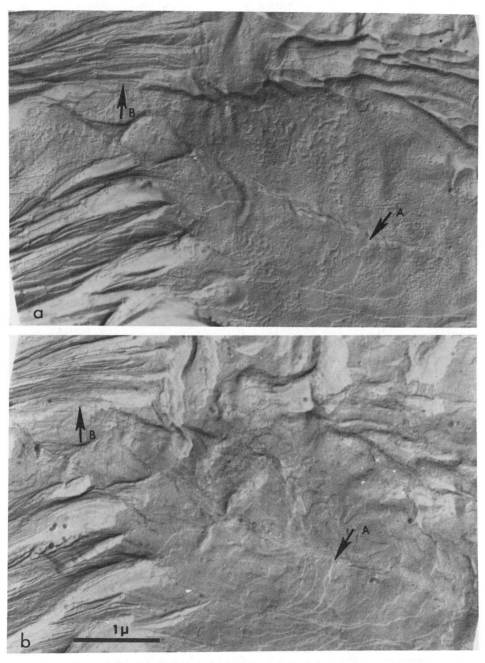

Figure 30   Electron micrographs of the same area of a free surface of POM-S2 obtained: a) before and b) after
anneal at 161°C.

Table 3

ℓ_1 and ℓ_2 During and After Annealing
for Various Periods at 159°C

Annealing Period	High Temp. Pattern ℓ_1 (A)	ℓ_2 (A)	Room Temp. Pattern ℓ_1 (A)	ℓ_2 (A)
No anneal	– –	– –	124	*
5 min.	215	– –	– –	– –
20 min.	– –	– –	164	58
1 hr.	210	– –	– –	– –
2 hr.	– –	52	170	50
7 hr.	215	52	173	62
17 hr.	215	55	167	71

NOTE:
   * peak not observed
   – – data not recorded

shows that considerable changes have also taken place in the observed morphological features. Although the gross features such as
spherulite boundaries and orientation of lamellar stacks remain
largely unchanged, the number of visible individual lamellae increased
significantly. The area indicated by arrow "A" was relatively free
of lamellar edges prior to the heating. After annealing approximately twice as many shadow lines indicating lamellar edges were
observed. In some cases (arrow "A") there appears to be a slight
suggestion of a change in structure (either a change from rough to
smooth surface or change in thickness) along the line at which a new
lamellar edge appears. In other cases lamellar edges appeared in
areas where there was no such precursor in the unannealed surface.
In some areas of the unannealed surface, where lamellae were in-
clined to the plane of the surface, the exposed edges frequently
grouped together to form large steps (arrow "B"). The corresponding
area of the annealed surface shows no such grouping of edges but
rather an array of regularly spaced lamellar edges all having approximately equal thickness.

Due to the difficulties of locating the same area on two separate replicas, latex particles were not used for shadow angle gauges;
hence no absolute step height measurement could be made from these
micrographs. However, since the surface was shadowed from the same
direction and elevation angle each time comparative measurements of
the lamellae thickness before and after annealing could be made.
Such measurements indicated no observable change in the step height
of those features which could be identified as individual lamellae
on the unannealed surface (e.g., arrow A). Comparisons could not
be made in areas, such as B, where several edges appeared to be

initially grouped together. The above described morphological
changes were observed at temperatures between 145°C and 161°C.
Annealing at and below 140°C produced no observable irreversible
changes in the surface structure or in the SAXD pattern.

In addition to observations of irreversible morphological
changes, investigations of modifications of the morphology corresponding to the reversible changes observed in the high temperature
SAXD patterns of polyoxymethylene were conducted. For this purpose
one of the samples (POM-S2) from the preceding series of experiments
was reheated to the previous annealing temperature of 157°C and
shadowed with Pt-C. Figures 31a and 31b are micrographs of a particular area on POM-S2 before and after annealing at 157°C for 5
hours. Figure 31c is the same area shadowed while the specimen was
held at 157°C. The SAXD patterns obtained from POM-S2 before,
during and after the first heating are given in Figure 32. All of
the morphological features observed in the annealed sample at room
temperature appeared unchanged in the replica which had been deposited at high temperature. Measurements of the lamellar edge shadows
indicated that if any changes in the step height occurred at high
temperature, they were less than 15%. On the other hand the reversible change in ℓ_1 was of the order of 30%. It should be noted that
the replica was removed at room temperature; however if substantial
changes in the surface occurred during cooling one would expect them
to be apparent as disruptions in the replica.

IR, DTA and Wide Angle X-Ray Measurements

Measurement of the temperature dependence of the SAXD pattern
of annealed poly(ethylene terephthalate) (PET-A) was made because of
the possibility of obtaining specific information concerning the
regular chain fold conformation at various temperatures through the
use of infrared spectroscopy. The SAXD data (Figure 33) indicate
that the long period increased slightly in a linear manner from -75°C
to just above the original annealing temperature of 180°C following
which a considerable increase occurred as the temperature was increased to a few degrees below the melting point (T_m=265°C). Upon
cooling, roughly one-half of the original increase in the long period
was recovered by 157°C. Further decrease in temperature resulted in
a slightly further decrease in long period along a curve which was
approximately parallel to the low temperature behavior of ℓ_1 during
the heating cycle. At room temperature an irreversible increase of
40% of the original room temperature ℓ_1 value was observed. The intensity of the scattering at the position of the discrete peak increased gradually and in a nearly reversible manner over the entire
temperature range (Figure 34). This increase became more pronounced
at temperatures above 180°C.

Using a different section of the PET-A specimen the infrared
absorbtion band at 988 cm^{-1}, which has been assigned to the regular

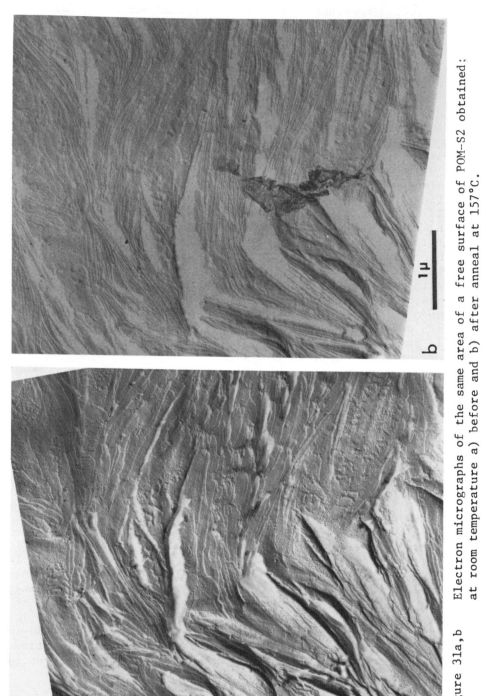

Figure 31a,b   Electron micrographs of the same area of a free surface of POM-S2 obtained; at room temperature a) before and b) after anneal at 157°C.

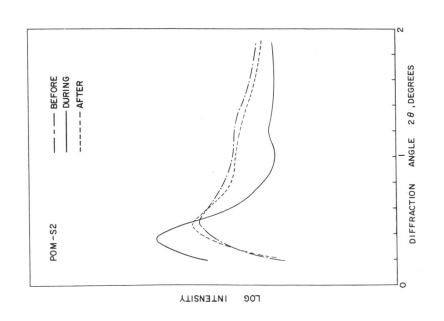

Figure 32   Small angle x-ray diffraction pattern from POM-S2 obtained before, during and after annealing at 157°C.

Figure 31c   Electron micrograph obtained at room temperature during reheating at 157°C. (From same sample as a and b.)

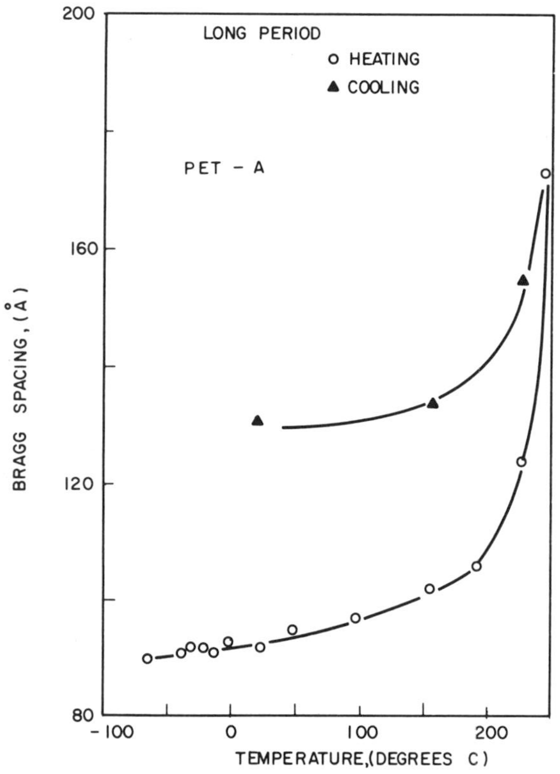

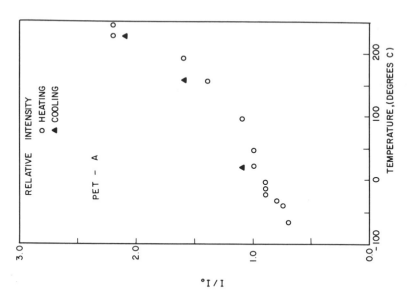

Figure 34 Plot of l_1 relative intensity as a function of temperature for PET-A.

Figure 33 Plot of l_1 as a function of temperature for PET-A.

fold structure (43), was measured between -75°C and 190°C. Representative IR spectra of this band at several temperatures are shown in Figure 35. Following Koenig and Hannon (43) the absorbance was calculated using the baseline indicated. The absorbance decreased linearly with temperature over the entire temperature range and extrapolated to zero at 240°C (Figure 36). In particular no effect is seen at 180°C. This is in reasonable agreement with the data of Hannon and Koenig (57) obtained on the same material after it had annealed at 240°C. The slope change at 40°C was probably an experimental artifact since, at this point, KBr windows were added and the cell evacuated for the high temperature measurements. Data could not be obtained above 198°C; it became impossible to resolve the absorption band.

Annealing studies using the dupont 900 Differential Thermal Analyzer (DTA) and measurements of wide angle x-ray diffraction line widths were also conducted for correlation with the annealing effects observed in the SAXD patterns. Figure 37 shows the melting thermograms obtained from POM-S2 which had previously been annealed at various temperatures for one hour in the DTA and cooled to room temperature. The arrows above the individual traces indicated the respective anneal temperatures. The melting thermogram of an unannealed sample is also included in Figure 37. The thermogram obtained from material which had been annealed at 138°C (or below) showed no significant differences from that of the unannealed material. Significant annealing processes occur at 142°C and higher.

The results of the wide angle line broadening measurements are given in Figure 38. The observed line width was corrected for instrumental factors and, assuming that crystalline disorder effects could be neglected, the crystallite size was calculated using the Scherrer (44) equation. This assumption is of course, not completely justified and subsequent discussion of the data will consider the effects of disorder. The crystallite size plotted in Figure 38 was calculated from the half width of 009 crystalline reflection from POM-Q3. Since the [009] direction of polyoxymethylene is normal to the basal plane of the lamellae this size corresponds to the thickness of the crystalline portion of the lamellae.

The crystallite size of the specimen before any annealing was 119 A while the SAXD long period ℓ_1 was 145 A. The temperature of the specimen was raised stepwise to 140°C. The total time that the specimen was held at each temperature was one hour, including a 30 minute period allowed for equilibration and completion of annealing processes. The sample temperature figures above 130°C are probably in error by $\pm$ 2°C due to the difficulties encountered in fixing the sample to the heat sink. The nature of the diffraction peak and the background were such that the probable error in determining the crystallite size was $\pm$ 5 A. The crystallite size began to increase between 100 and 120°, becoming 145 A at 140°C.

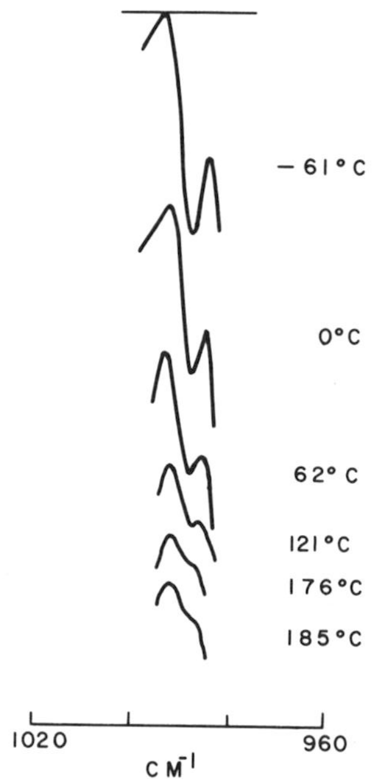

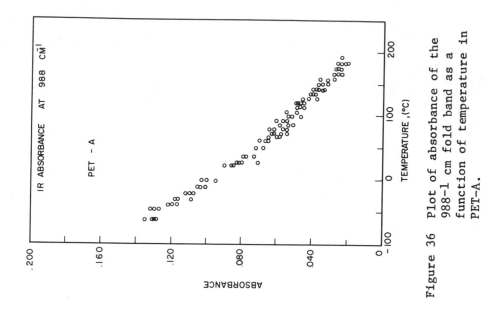

IR ABSORBANCE AT 988 CM⁻¹

PET - A

Figure 36 Plot of absorbance of the
988-1 cm fold band as a
function of temperature in
PET-A.

Figure 35 Infrared 988 cm⁻¹ absorbtion
band of PET-A at various
temperatures.

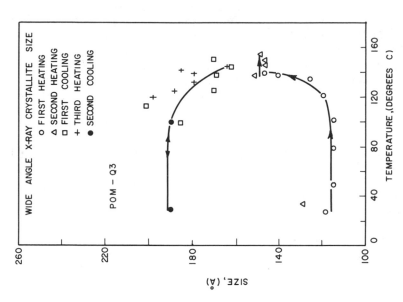

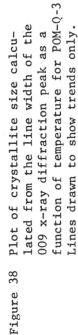

Figure 38   Plot of crystallite size calculated from the line width of the
009 x-ray diffraction peak as a
function of temperature for POM-Q-3
Lines drawn to show trends only.

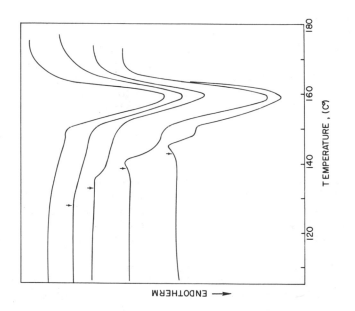

Figure 37   DTA melting thermograms of POM-S2
after annealing at various temperatures. The annealing temperature
is indicated by the arrow placed
above the individual scans.

At this point in the experiment, due to sample mounting problems, it became necessary to cool the specimen to near room temperature. The crystallite size was found to have decreased to 130 A at 36°C. The measurements were resumed by heating immediately to 138°C where a size of 153 A was measured. Further heating to 155°C caused no measurable change. Cooling from this temperature to 100°C in steps resulted in a further <u>increase</u> in crystallite size to 187 A. Re-heating from 100°C to 145°C caused a decrease in crystallite size along the line generated by the data of the previous cooling measurements. Upon return to room temperature the crystallite size was found to be the same as that of 100°C, i.e., 187 A. The SAXD ℓ_1 and ℓ_2 values at room temperature were 176 A and 77 respectively.

Lattice distortions and defects, if present, also contribute to the breadth of the wide angle x-ray diffraction peaks. These effects can, in principle, be separated by comparison of the widths of higher order diffraction peaks (45). In the present case this was not possible since higher orders of the 009 peak were not de-tected. The good agreement between the room temperature crystallite size measurements and the ℓ_1 values indicates that axial broadening of the 009 reflection due to disorder effects was small compared to that due to the crystallite size effects. In subsequent measurement of the SAXD pattern of this specimen at elevated temperature ℓ_1 in-creased reversibly from 176 A to 189 A in the 100°C to 155°C temper-ature range.

SUMMARY AND CONCLUSIONS

A wide variety of results obtained from several different types of experiments have been presented and discussed at some length in the preceding section. In the introduction the efforts of others, both experimental and theoretical, to interpret SAXD from crystal-line polymers were briefly reviewed. In this section the current data will be examined in terms of previously proposed models and theories of diffraction. Most of the diffraction results are found to be consistent with Fischer's models of premelting (58,61).

To facilitate this discussion the major observations of this work are summarized in the list below. The first 3 items deal with EM-SAXD measurements on the same specimens at room temperature with the remainder dealing with reversible and irreversible effects of annealing on the SAXD pattern and other properties. This division corresponds to two of the three problem areas outlined in the Intro-duction. The third area, the ℓ_1/ℓ_2 ratio, is of a general nature and observations pertaining to it are contained in both sections.

1. SAXD measurements on unoriented polyethylene (LPE-Q) and poly-oxymethylene rapidly cooled (POM-Q1) and slowly cooled (POM-S1) show one strong discrete peak at scattering angles between 0.2 and 1.0 degrees (ℓ_2) and, usually, one additional very weak

peak (ℓ_2) at wider angles.
a. Corrected ℓ_1/ℓ_2 ratios range from 2.3 to 2.7
b. Surface EM replicas show lamellar step heights which agree
    reasonably well with ℓ_2.

2. Fracture surfaces of isothermally crystallized POM-I show two
    types of structure:
    a. Lamellae oriented parallel to the fracture surface having
        step heights of 140 to 160 A
    b. Spherical particles arranged on the fracture surface in a
        manner which suggests that they represent lamellae oriented
        perpendicular to the surface. The thickness of the under-
        lying structure, indicated by the spacing of the particle
        rows, is 220 to 320 A.
    SAXD information indicates two major physical repeat distances
    (170 and 310 A from the correlation function). The ℓ_1/ℓ_2 ratio
    from the corrected diffraction pattern was 1.6.

3. EM measurements of oriented polyoxymethylene (POM-D1), using
    gold decoration, indicate lamellae oriented perpendicular to
    the surface with a thickness of 100 A to 130 A. SAXD measure-
    ments on the same material show two long periods, 155 A and
    62 A, giving an ℓ_1/ℓ_2 ratio of 2.5 while correlation function
    calculations yield a strong peak at 148 A and a weaker 2nd
    neighbor peak at 324 A.

4. For a POM sample crystallized by slow cooling (POM-S2):
    a. ℓ_1 (165 A) is nearly independent of specimen temperature
        from room temperature up to 134°C. From this temperature
        to 161°C ($T_m-2°$), ℓ_1 increases to 195 A. The same curve
        was followed upon cooling to room temperature and reheating.
    b. The intensity of the ℓ_1 peak reversibly followed a curve
        which increased continuously from room temperature to a
        maximum at 150°C, decreasing slightly above that tempera-
        ture; then the ℓ_1 spacing and intensity change was com-
        pletely reversible.
    c. ℓ_2 remained constant at 70 A while the specimen tempera-
        ture was raised to 161°C and then reduced to 150°C over a
        period of about 20 hours. Further cooling resulted in a
        step change in ℓ_2 to a value of 90 A at 124°C and a con-
        siderable enhancement of the intensity. After this treat-
        ment further cooling and reheating produced continuous,
        reversible changes in ℓ_2 similar to those observed for ℓ_1.

5. Annealing of POM-Q2, which had been rapidly cooled from the
    melt, produced results similar to the above except for an
    initial irreversible increase of ℓ_1 that occurred in the 130 to
    150°C temperature range. The initial room temperature value of
    ℓ_1 (145 A) increased only slightly (to 155 A) at 128°C. Above
    this temperature it increased rapidly, reaching 191 A at 151°C,

and continuing to 210 A at 161°C. During this treatment the
ℓ_1 intensity changed in a manner similar to that seen for POM-S2
(item 4b). Subsequent heating and cooling cycles produced
reversible spacing (165-210 A) and intensity changes. ℓ_2 was
seen to increase from 55 A at room temperature to 60 A at 161°C.
Upon cooling, ℓ_2 behaved as described in item 4c.

6.   The reversible as well as the irreversible changes in ℓ_1 occurred
     within 5 minutes (the minimum experimental time possible) of
     being heated to 159°C.

7.   SAXD measurements at elevated temperatures on the oriented and
     briefly annealed specimen (POM-D) produced results similar to
     those observed for POM-S2 (item 4 above).

8.   The ℓ_1/ℓ_2 ratio in all three of the above series of measurements
     (POM-S2,-D, and -Q2) was initially between 2.35 and 2.5, in-
     creased to about 3.0 at temperatures near the melting point (due
     to the increase in ℓ_1 while ℓ_2 remained nearly constant) and
     decreased abruptly to 1.9 to 2.1 after the change in ℓ_2 (item 4c).

9.   EM observations of a particular area of a slowly cooled poly-
     oxymethylene (POM-S2) free surface before and after annealing
     show that:
     a.   No observable change in morphology occurred during anneal-
          ing or was observed after annealing at or below 140°C
     b.   In areas where the lamellar were nearly parallel to the sur-
          face and where the edges appeared well spaced, the number
          of lamellae present, determined by the number of edges seen,
          appeared to approximately double upon annealing at 145°C or
          higher.
     c.   Annealing at 145°C or higher caused a reorganization of the
          surface in regions where the lamellae were originally in-
          clined to the surface. This caused lamellae edges, which
          had been originally indistinct or grouped together, to be-
          come clearly visible as distinct individual edges well
          separated from the adjacent edges. New lamellar edges may
          also have appeared in these areas due to the doubling phen-
          omenon described in 9b above.
     d.   No observable change in the lamellar thickness occurred for
          any annealing temperature although the SAXD ℓ_2 increased
          irreversibly by a considerable amount during cooling from
          the higher anneal temperatures.

10. Comparisons of lamellar thickness measurements made from a
     replica deposited on a specimen at 157°C, after it had been
     annealed, with measurements of the same lamellae at room temper-
     ature after the initial anneal treatment, showed no measure-
     able change. SAXD patterns obtained at this temperature showed
     a 25% reversible increase in ℓ_2. The number and arrangement of

observable lamellae also remained the same.

11. DTA annealing and experiments indicated that significant anneal-
     ing occurred at temperatures of 142°C and above. Below that
     temperature no annealing effects were observed.

12. Measurements of the crystallite size in the c axis direction
     during annealing, by means of wide angle x-ray diffraction line
     widths, on a rapidly cooled POM specimen (POM-Q3) showed that
     the original size (120 A) was maintained up to a temperature of
     135°C. Further increase in specimen temperature resulted in an
     irreversible increase in the crystallite size to 155 A at 155°C.
     Cooling from that point resulted in a <u>reversible increase</u> (dur-
     ing subsequent cycles) in crystallite size to 190 A at 120°C.
     Further cooling to room temperature produced no further change.
     The initial SAXD ℓ_1 value of POM-Q3 prior to the above treat-
     ment was 158 A. After this initial anneal during the wide angle
     x-ray measurement, the temperature characteristics of ℓ_1 were
     found to be similar to that of POM-S2, showing no significant
     change between room temperature and 100°C and increasing rever-
     sibly from 176 A at 100°C to 188 A at 155°C.

13. Infrared measurements of PET-A at temperatures between -70°C
     and 245°C reveal a uniform decrease in the fold band absorbance
     with increasing specimen temperature, while SAXD measurements
     over the same temperatures range show:
     a. Both reversible and irreversible changes in ℓ_1 similar to
         those noted in item 5 for POM
     b. A generally linear increase in ℓ_1 intensity with specimen
         temperature from -70°C to 100°C. Above 100°C the tempera-
         ture dependence of the ℓ_1 intensity increased considerably.

14. The SAXD pattern produced by nylon-11 (N-11-S) was observed to
     undergo both reversible and irreversible changes with specimen
     temperature. The long period increased initially in two steps
     of about 17 A at 135°C and 160°C. During cooling and subse-
     quent temperature cycles reversible changes in long period
     again occurred discontinuously at 160°C. Below 160°C ℓ_1 de-
     creased in a continuous manner by 12 A. The long period ob-
     served at room temperature after annealing was somewhat larger
     than the initial long period.

     In the following discussions of these observations in terms of
models and conclusions, specific items of the above list will be
referred to by a number in order to avoid repitition of the description of a specific observations. We consider first the x-ray
results by themselves, asking whether they can be interpreted in a
self consistent manner and later consider the problem of correlating them with the EM results.

One of the problems stated in the introduction concerned the self consistency of the observed SAXD pattern. Two or more discrete peaks have commonly been observed in positions which were not consistent with a single physical repeat distance, i.e., the ratio ℓ_1/ℓ_2 departed from 2. Some authors have suggested that proper treatment and correction of the data would reduce the ratio to 2 (11,12). The current results (item 1a) show that this is not the case for as-crystallized, bulk polyethylene and polyoxymethylene. While corrections did alter the ratios (see Table 2) they did not ultimately approach a value of 2 or for that matter any common value, integer or non integer.

Reinhold, et al., have suggested (34) that the ratio departs from integer values because of the disordered nature of the diffracting structure and that the patterns should be interpreted by application of paracrystal diffraction theory (59). They find that ratios deviating from integer values can be obtained from systems having a single physical repeat structure in which the individual distances are given by an asymmetrical distribution function. Applying this method to the present observations (items 1 to 3), we find values of the asymmetry parameter (γ) of -0.32 for POM-I and of 0.35 to 0.50 for LPE-Q and POM-Q1, D and S1. To our knowledge no values for γ have been reported by others; a positive γ is said to be expected due to increases in thickness of lamellae during crystallization (34). The definition of γ is such that the position of the maximum in the lamellar spacing distribution function is displaced from the average spacing by a factor of $(1-\gamma)$. This means that, in the case of POM-I, the spacing of the most numerous lamellae would be 1.3 times larger than the average spacing. For the other specimens this value would be less than the average spacing by a factor of 0.50 to 0.65.

Finally, we must consider the possibility that the two SAXD peaks as seen in the as-crystallized material are the first order diffraction peaks from two distinct structures within the specimen. Items 4c and 7 show that changes in position and intensity of the two peaks with increasing temperature occurred independently. If they were related through models such as that of Reinhold, et al., (34) one would expect these changes in the peaks to be coordinated in some way.

The apparent abrupt shift of the ℓ_2 peak to a position such that ℓ_1/ℓ_2 nearly equaled 2.0 upon cooling, is believed to be not a shift of the original ℓ_2 but rather the appearance of a new peak which is actually the second Bragg order of the ℓ_1 peak. The appearance of this new peak, which was significantly more intense then the original ℓ_2, may obscure the original peak so that it becomes undetectable. The reason for the sudden appearance of the new second order peak is probably an increase in the periodic nature of the ℓ_1 physical repeat distance. In terms of paracrystal diffraction theory the

presence of two orders requires that the standard deviation of a
Gaussian lattice distribution function, divided by the average lattice distance, fall below a value of 0.2. Although this increase
in order is opposite that found during the annealing of polyethylene
single crystal mats, in which the number of higher order reflections
observed generally decrease (60), it is not unreasonable to expect
it to occur in the bulk.

The reversible increase of the ℓ_1 intensity followed by a decrease just below the melting temperature (item 6) has been explained by Fischer et al., (13,61) in terms of a surface premelting model.
In order to account for the increase in long period with increasing
specimen temperature, Fischer modified the original premelting model
(58). In its original form the model postulated a surface melting
of the crystallites which increased the amount of material in the
amorphous phase at the expense of the material in the crystalline
phase. The only change in the overall repeat distance (amorphous
plus crystalline phase) was that accompanying the change in specific
volume of the portion "melting". The modification involved the
assumption that the premelting of the earlier model is accompanied
simultaneously by a translation of the chains within the crystalline
phase such that the thickness of that phase remains constant; hence,
the overall repeat distance can increase considerably. As predicted
by these premelting models, the intensities of the ℓ_1 peaks of all
POM specimens as well as the PET-A sample increased during heating,
corresponding to the conversion of material from the crystalline to
the amorphous phase. Above a temperature at which the material is
equally divided between the two phases, the ℓ_1 intensity should decrease.

Reversible increases in ℓ_1 with increasing temperature were
observed in all specimens measured. The 25% shift in POM (Items 5
and 10) is considerably less than the shifts reported by O'Leary and
Geil (16). However, long periods calculated from our uncorrected
SAXD patterns showed reversible shifts of about the same amount as
they reported. Since their patterns were not corrected for slit
smearing and Lorentz-geometric effects it is believed that this is
the difference between the two results. It is interesting to note
that changes in long period can be predicted on the basis of the
original Fischer model. This is due to the fact that conversion of
material from the crystalline to the amorphous phase causes an increase in the specific volume of the material. Since the periodicity
for SAXD purposes consists of both phases, the net result is an increase in the periodicity proportional to the amount of volume
change (assuming a one dimensional system).

In an effort to determine the cause of the changes of long
period with temperature, the dimensions of the crystallite were
determined as a function of temperature by means of wide angle x-ray
line broadening measurements (item 13). After an initial increase,

which seems to correspond to the irreversible increase in ℓ_1 (item 5), the crystallite size was observed to <u>decrease</u> reversibly while the ℓ_1 intensity and period <u>increased</u> reversibly. In converting from the crystalline to the amorphous state the specific volume of polyoxymethylene increases from 0.662 cc/gm to 0.905 cc/gm (measured at room temperature) (62). This means that the material involved in the 45 A decrease in crystallite size during heating from 100°C to 150°C must appear in the inter-lamellar region as 55 A of amorphous material, an increase of 10 A. The change in ℓ_1 over this temperature range for the same material was 12 A.

As previously pointed out, the premelting theory predicts an increase in the intensity of the diffraction as the temperature is increased to the point where the material is equally divided (with respect to weight) between the two phases; further heating causes a decrease in intensity. Using the above specific volume figures and assuming that less than 5% of the material is in the amorphous phase at room temperature, application of the original Fischer model predicts a long period increase of 11% between room temperature and the room temperature at which the intensity maxima is observed. Data obtained from all of the POM samples which were measured at elevated temperature indicate increases of between 17 and 18% at the temperature at which the intensity maxima was observed. The specific volume figures used above represent room temperature conditions. Increased temperature would of course cause both figures to increase due to thermal expansion. Since the crystalline phase is the more tightly bound structure it is reasonable to expect that the amorphous specific volume would show the larger increase. Increases of 5% and 15% for the crystalline and amorphous specific volumes at 155°C are reasonable estimates based on Swan's work with polyethylene (63) and would be sufficient to cause the predicted increase in long period to agree with the observed 17% increase. Aoki et al., (19) using SAXD measured values of thermal expansion, indicate values of 0.68 and 0.88 cc/gm for the crystalline and amorphous specific volumes of 159°C, i.e., increases of 3% and 10% respectively. Further, since the premelting process would be expected to be essentially reversible with temperature, the predicted change in long period would likewise be reversible.

For polyoxymethylene the onset of the SAXD and wide angle line width irreversible annealing processes (Figures 17 and 38) and the sharp upturn in the ℓ_1 vs. T curves for annealed and slow cooled POM (Figures 21 and 13) occur between 140 and 145°C. This is also the temperature range at which DTA measurements indicate the onset of significant annealing processes. It should be noted here that McCrum (64) and Takayanagi (65) have reported an α peak in the dynamic mechanical loss spectrum of polyoxymethylene in the 130-150°C range. This loss peak is thought to be associated with the crystalline phase (66) and in other polymers has been suggested to arise with the onset of rotation and translation of segments within the crystal (67).

Attempts to correlate the changes in the ℓ_1 intensity with
specific changes in the fold conformation of the polymer molecule by
infrared techniques were inconclusive (item 13). On the basis of
the Fischer model, at the temperature at which premelting begins, a
change in the slope of the gradual decrease in the fold band absorbance would be expected. No such change in slope was observed but
the band became unresolvable at a temperature near that at which ℓ_1
began to increase significantly. It does appear, however, that given
an IR fold band which was either more isolated or stronger than the
988 cm^{-1} band in PET, the experiment could yield results which
would indicate whether or not such a premelting accompanied the ℓ_1
intensity change.

The studies regarding the changes in the SAXD pattern as a
function of specimen temperature have shown that, with the notable
exception of nylon-11, the primary long period of the materials
examined was a continuous function of the specimen temperature. The
nylon-11 results appear somewhat anomalous in that the long period
was observed to change in a discontinuous manner at definite temperatures. Discontinuous irreversible changes in the long period of
various nylon single crystal mats during annealing have been obser-
ved(55). In the initial observations a doubling of the crystalline
lamellar thickness was proposed to occur by a mechanism involving
the passage of individual folds through the body of the lamella
until they appear as long loops on the opposite side. This mechanism results in a doubling of the thickness with hydrogen bonds along
a given segment being broken simultaneously to permit the molecular
translation as a segmental unit. It thus appears that these secondary bonds exert a strong force which stabilizes the fold period of
the lamellae against the tendency to increase in length during
annealing.

Another factor which would affect the thickening of a lamella
consisting of regularly folded molecules is the length of the unit
cell along the chain axis (55). Regular folding implies a unique
conformation of the molecule at the fold meaning that any change in
the location of the fold along the chain would necessarily be restricted to integer multiples of the c axis repeat distance (with
irregular folding step changes in crystal thickness but not necessarily lamella thickness would be anticipated). In materials having
a short repeat distance, such as polyethylene, this effect would
produce step changes so small that they would appear as a continuous
change in SAXD experiments. Materials with longer repeat distances
such as nylon-11 and POM would be expected to show detectable step-
wise changes in their long period if the above conditions regarding
regular folding were met. The current results indicate that discontinuous changes in the nylon-11 long period do proceed in steps
which are approximately equal to that material's c axis repeat dis-

tance of 15 A (68). Although this is not observed in the polyoxymethylene results, the six fold symmetry about the c axis coupled
with molecular rotation may permit smaller steps. Since the original
report of a doubling in period for nylon single crystal mats, similar
step-wise increases in spacing by single or multiple c axis repeats
have been described for a number of nylon single crystal mats (55b).

Considering these results in terms of Fischer's models, we see
that while step-wise increases nearly equal to the unit cell c distance could not be satisfactorily explained on the basis of the original model, the modified premelting concept which requires simultaneous melting and annealing processes is not completely ruled out.
Motion of the molecule along the chain axis would presumably be
subject to the above described restrictions during annealing, each
repeat unit "melting" contributing an increase in local thickness
of about 1/2 a repeat distance (corrected for amorphous/crystal
specific volume).

If one accepts the conclusion that the increase in the nylon-11
long period with increasing sample temperature is due to an increase
in the fold period then one must also consider the step-wise decrease in the long period upon cooling. A decrease in crystal thickness is not in accord with the kinetic theory of crystallization (69)
which predicts that the most stable crystallized state is that of
extended chains and that once increased by annealing, the fold would
not be expected to decrease at any temperature. However, in addition to the thermodynamic theory of crystallization (70) which does
predict a possible decrease in crystallite thickness, Lindenmeyer
(71) has recently suggested that several metastable states may exist
at segment distances less than the length of the entire molecule.
The length of these segments was predicted to change directly and
reversibly with temperature.

On the basis of the above results we can conclude that in all
of the as-crystallized material, where two discrete peaks were observed in the SAXD pattern, they represented two separate physical
repeating structures, and that changes occurring in l_2 at elevated
sample temperature can be explained as an increase in the regularity
of the l_1 repeat structure; a new l_2 being observed. The reversible
temperature of l_1 in POM both in terms of intensity and long period,
at least up to the maximum temperature used for the WAXD line width
measurements, can be reasonably explained in terms of a premelting
of the surface of the crystalline phase such as in the original
Fischer model. On the other hand nylon-11 exhibits a quite different
behavior. The step-wise change in long period together with the
essentially temperature independence of intensity of the peak leads
us to discard a premelting explanation here. Instead, the change in
long period may be due to a reversible change in the thickness of
the crystalline layer with little change in the amorphous layer, i.e.,
a regular fold structure is retained and stabilized by H-bonding.

In addition the importance of applying not only slit corrections, but also the Lorentz-geometric correction to the SAXD data is shown.

The situation regarding the interpretation of the EM results remains unclear. We have shown that in PE and POM the thickness of lamellae observed on a free surface corresponds to the very weak ℓ_2 SAXD peak, and that structures having dimensions corresponding to the dominate ℓ_1 peak were seen only in areas of fracture surfaces in which lamellae appear to be oriented more or less perpendicular to the surface. This, along with the previously mentioned results of Hase (53), supports the conclusion stated earlier that ℓ_1 and ℓ_2 may represent different physical structures within the material. The reason that the ℓ_1 structure is generally not observed on free surfaces and the relationship between ℓ_1 and ℓ_2 is at present unknown. Since lamellae on the free surface of a melt crystallized specimen crystallize at lower temperatures than the bulk of the specimen, they would be expected to be thinner than those located in the interior. However, these thinner lamellae are also observed in the interior of the specimen and further, there is no reason to expect that two distinct structures would be formed in this way.

Observations concerning changes in the arrangement of the lamellae and the apparent increase in number of lamellae on a free surface upon annealing have indicated that the process is temperature dependent only to the extent that a threshold value must be exceeded for any change to be observed. Annealing at 161°C produced the same results as treatments of 145°C. The effect of time at the anneal temperature was not investigated. Thus we are left with the problem of suggesting a relationship between the ℓ_1 and ℓ_2 structures. Other than for an apparent doubling of lamellae, suggesting some form of pairs of lamellae, seen frequently, but not always, on the repeated replica surface, there seems no other clue at present. However, we can suggest no reasonable explanation why a pair of lamellae, presumably formed by interlamellar material, should delaminate when fractured parallel to the lamellar surface, but remain as a single unit when fractured normal to the surface. It is hoped that future work in our laboratories on poly-4-methyl-pentene-1, which has ℓ_1 lamellae ca. 400 A thick and therefore readily visible, but which, unfortunately, yield very poor SAXD patterns, may contribute to solving this problem. Likewise further study of the annealing of various nylons by SAXD and wide angle x-ray diffraction line broadening may help clarify the cause of the step changes in spacing.

In terms of the three problems cited in the introduction the situation must still be considered unsatisfactory. Considering SAXD data only; (a) for the experimental conditions used (maximum temperature and time of annealing) the reversible change in intensity and ℓ_1 spacing for POM can be adequately interpreted in terms of a lamellar surface premelting model and (b) after data correction $\ell_1 = 2\ell_2$ for annealed POM and LPE(1) samples. However, premelting does not appear to satisfactorily explain the reversible change in ℓ_1 spacing for nylon-11 and, in particular, $\ell_1 \neq \ell_2$ even after data correction for as-crystallized POM and LPE. The situation is further worsened by comparison of the EM and SAXD data. Although for LPE after annealing, ℓ_1 agrees with EM observations (1), for LPE prior to annealing and for POM both before and after annealing, ℓ_2 agrees most closely with EM observed step heights. Only on those portions of fracture surfaces in which the lamellae are perpendicular to the surface, is a periodic structure seen with a spacing in agreement with ℓ_1. Not only does this still raise questions concerning the general interpretation of SAXD (and EM) data from crystalline polymers in terms of lamellar thickness, but also makes one question the apparent adequate interpretation of reversible changes listed above.

ACKNOWLEDGEMENT

This research was supported by the U. S. Army Research Office, Durham.

REFERENCES

1. Geil, P.H., Polymer Single Crystals, Interscience, New York(1963)
2. Brown, R.G., and R.K. Eby, J. Appl. Phys., 35, 1156 (1964).
3. Blais, J.J.B.P., and R. St. John Manley, J. Macromol. Sci.(Phys.) B1, 525 (1967).
4. Illers, K.H., and H. Hendus, Kolloid-Z.u.Z. Polymere, 218, 56 (1967).
5. McHugh, A.J., and J.M. Schultz, Phil. Mag., 24, 155 (1971).
6. Fischer, E.W., H. Goddar and G.F. Schmidt, Kolloid-Z.u.Z. Polymere, 226, 30 (1968).
7. Luch, D., and G.S.Y. Yeh, J. Macromol. Sci.(Phys.), B7, in press.
8. Hendus, H., Ergeb. Exakt. Naturwiss., 31, 331 (1959).
9. Schultz, J.M., W.H. Robinson and G.M. Pound, J. Polymer Sci., A2, 5, 511 (1967).
10. Hoffman, J.D., and J.J. Weeks, J. Chem. Phys., 42, 4301 (1965).
11. Kortleve, G., and C.G. Vonk, Kolloid-Z.u.Z. Polymere, 225, 124 (1968).
12. Kavesh, S., and J.M. Schultz, J. Polym. Sci., A2, 9, 85 (1971).
13. Nukushina, Y., Y. Itoh and E.W. Fischer, Polymer Letters, 3, 383 (1965).
14. Nobuta, A., A. Chiba, and M. Kaneko, Rept. Prog. Polymer Phys., Japan, 12, 137 (1969).
15. Tsvankin, D.Y., and Yu. A. Zubov, Vysokomol. Soedin., 7, 1848 (1965).
16. O'Leary, K., and P.H. Geil, J. Macromol. Sci.(Phys.), B1, 147 (1967).

17. Dawkins, J.V., P.J. Holdsworth and A. Keller, Makromol. Chem.,
    118, 361 (1968).
18. Fischer, E.W., Pure Appl. Chem., 26, 385 (1971).
19. Aoki, Y., A. Nobuta, A. Chiba and M. Kaneko, Poly. J., 2, 502
    (1971).
20. Maeda, M., K. Miyasaka and K. Ishikawa, Kobunki Kagaku, 26,
    241 (1969).
21. Fichera, A., C. Garbuglio and R. Zannetti, Chemie Ind., 52,
    983 (1970).
22. Krigbaum. W.B., Y.I. Balta and G.H. Via, Polymer, 7, 61 (1966).
23. Geil, P.H., J. Polymer Sci., Pt. C, 13, 149 (1966).
24. Guinier, A., and G. Fournet, Small Angle Scattering of X-Rays,
    John Wiley, New York, 1955.
25. Beeman, W.W., P. Kaesberg, J.W. Anderegg, and M.B. Webb, Hand-
    buch der Physik, 32, 191 (1967).
26. Lake, J.A., Acta Cryst., 23, 191 (1967).
27. Kijkstra, A., G. Kortleve and C.G. Vonk, Kolloid-Z.u.Z. Polymere,
    210, 121 (1966).
28. Ruland, W., Acta Cryst., 17, 138 (1964).
29. Stroble, G.R., Acta Cryst., A26, 367 (1970).
30. Schmidt, P.W., and R. Hight, Jr., Acta Cryst., 13, 480 (1960).
31. Alexander, L.E., X-Ray Diffraction Methods in Polymer Science,
    Wiley-Interscience, New York, 1969.
32. Gella, R.J., B. Lee, and R.E. Hughs, Acta Cryst., A26, 118 (1970).
33. Hoseman, R., Z.Phys., 128, 1 and 465 (1950).
34. Reinhold, Chr., E.W. Fischer and A. Peterlin, J. Appl. Phys.,
    35, 71 (1964).
35. Kawai, T., M. Hosoi and K. Kamide, Makromol. Chem., 146, 55 (1971).
36. Blundell, D.J., Acta Cryst., A26, 472, 476 (1970).
37. Tsvankin, D. Ya., Vysokomol. Soedin., 6, 2078 (1964).
38. Tsvankin, D. Ya., Vysokomol. Soedin., 6, 2083 (1964).
39. Blasie, J.K., and C.R. Worthington, J. Mol. Biol., 39, 417 (1969).
40. Debye, P., H.R. Anderson, Jr., and H. Brumberger, J. Appl. Phys.,
    28, 679 (1957).
41. Vonk, C.G., and G, Kortleve, Kolloid-Z.u.Z. Polymere, 220, 19
    (1967).
42. Kortleve, G., and C.G. Vonk, Kolloid-Z.u.Z. Polymere, 225, 124,
    (1968).
43. a) Koenig, J.L. and M.J. Hannon, J. Macromol. Sci.(Phys.), B1,
    119 (1967)(PET); (b) Koenig. J.L., and M.C. Agboatwalla, J.
    Macromol. Sci.(Phys.), B2, 391 (1968); Frayer, P.D., J.L. Koenig
    and J.B. Lando, J. Macromol. Sci.(Phys.), B3, 329 (1969)(Nylon
    66); (c) Frayer, P.D., J.L. Koenig, J.B. Lando and B. Hickel,
    J. Macromol. Sci.(Phys.), B6, 129 (1972)(Nylon 6); (d) Koenig,
    J.L. and P.D. Vasko, J. Macromol. Sci.(Phys.), B4, 347 (1970)
    (Amylose).
44. Klug, H.P. and L.E. Alexander, X-Ray Diffraction Procedures,
    Wiley, New York, 1954.

45. Bonart, R., R. Hosemann and R.L. McCullough, Polymer, 4, 199
    (1963).
46. Statton, W.O., J. Polymer Sci., 41, 143 (1959).
47. Burmester, A., Ph.D. Thesis, Case Western Reserve University
    1970.
48. Perrell, D., M.S. Thesis, Case Western Reserve University,1968.
49. Spit, B.J., J. Macromol. Sci.(Phys.), B2, 45 (1968).
50. Keller, A., and M. Machin, private communication (1969).
51. Siegmann, A., and P.H. Geil, J. Macromol. Sci.(Phys.), B4, 557
    (1970).
52. Geil, P.H., J. Polymer Sci., 47, 65 (1960).
53. Hase, Y., and P.H. Geil, Polymer J., 2, 560 (1971).
54. Peterlin, A., and K. Sakoaku, J. Appl. Phys., 38, 4152 (1967).
55. a) Dreyfuss, P., and A. Keller, J. Macromol. Sci.(Phys.), B4,
    811 (1970) and J. Polymer Sci., B8, 253 (1970); (b) Burmester,
    A.F., P. Dreyfus, P.H. Geil and A. Keller, in press.
56. Spegt, P.A., Makromol. Chem., 139, 139 (1970).
57. Hannon, J.J., and J.L. Koenig, J. Polymer Sci., A2, 7, 1085
    (1969).
58. Fischer, E.W., Kolloid-Z.u.Z. Polymere, 231, 458 (1969).
59. Hoseman, R., and S.N. Bagchi, Direct Analysis of Diffraction by
    Matter, North Holland, Amsterdam, 1962.
60. Statton, W.O., and P.H. Geil, J. Appl. Polymer Sci., 3, 357
    (1960).
61. Fischer, E.W., Kolloid-Z.u.Z. Polymere, 218, 97 (1967); H. Goddar,
    G.F. Schmidt and E.W. Fischer, Makromol. Chem., 127, 286
    (1969).
62. Hammer, C.F., T.A. Koch, and J.F. Whitnery, J. Appl. Polymer Sci.,
    1, 169 (1959).
63. Swan, P.R., J. Polymer Sci., 42, 525 (1960).
64. McCrum, N.G., J. Polymer Sci., 54, 561 (1961).
65. Takayanagi, M., and S. Minami, Rept. Progr. Polymer Phys.(Japan),
    7, 241 (1964).
66. McCrum, N.G., B.E. Read and G. Williams, Anelastic and Dielec-
    tric Effects in Polymeric Solids, John Wiley and Son, New
    York (1967).
67. Takayanagi, M., Mem. Fac. Eng. Kyushu Univ., 23, 41 (1963).
68. Miller, R.L., and L.E. Nielson, J. Polymer Sci., 44, 391 (1960).
69. Hoffman, J.S., SPE Trans., 4, 315 (1964).
70. Peterlin, A., and E.W. Fischer, Z. Physik, 159, 272 (1960), and
    A. Peterlin, E.W. Fischer and Chr. Reinhold, J. Chem. Phys.,
    37, 1403 (1962).
71. Lindenmeyer, P.H., J. Chem. Phys., 46, 1902 (1967).

CHAIN FOLDING AND THE LOCATION OF THE AMORPHOUS COMPONENT IN POLY(TRANS 1,4 BUTADIENE) SINGLE CRYSTALS

A. E. Woodward, S-B. Ng and J. M. Stellman

Department of Chemistry, The City College of

The City University of New York, New York, 10031

ABSTRACT

The preparation of uniform monolayer crystals of poly (trans-1, 4 butadiene)-PTBD-crystals by a "minimum dissolution" temperature techniques is described. Epoxidation in benzene suspension of PTBD crystals grown from various solvents shows that the fraction of double bonds available for reaction varies with the growth solvent and polymer molecular weight. The number of monomer units per chain fold is calculated from the epoxidation results, the chain repeat distance and the crystal thickness and is found to depend upon the growth solvent. Infrared spectra, differential scanning calorimetry and broad-line n.m.r. spectra show that the total amorphous fraction in PTBD crystals grown from some media substantially exceeds the fraction of material at the crystal surfaces available for epoxidation. The results of preliminary solvent dissolution studies are also reported.

INTRODUCTION

Prior to the inception of this work in 1968 almost all experimental investigations of chain folding in polymer single crystals had been carried out on polyethylene. Considerable progress was made by Keller, Peterlin, Flory, Madelkern, Fischer and coworkers before and after 1968 towards elucidation of chain folding and the location of the amorphous component in polyethylene crystals. However, one problem has always been present. That is the number of monomer units on the surfaces of polyethylene single crystals can not be determined in a direct quantitative

101

fashion. Most of the physical methods employed, such as infrared
spectroscopy, calorimetric determinations and density measurements
yield an "amorphous" fraction but do not specify where this
amorphous fraction resides in the crystal.[1] By studying the
intensities of low angle X-ray reflection as a function of temperature, Fischer and coworkers have shown that there is an amorphous
layer on the surfaces of polyethylene [2] and polybutene-1 [3]
single crystals. From early   chemical studies of a destructive
nature e.g. reaction with fuming nitric acid, a similar conclusion was reached for polyethylene single crystals.[4, 5] More
recent results show this conclusion to be erronous.[6,7] However,
in spite of the large number of investigations carried out on
polyethylene a quantitative estimate of the fraction of amorphous
material residing on the crystal surfaces has not been given.

    With this in mind a search was instituted for a crystallizable
polymer which had chemically detectable groups in the repeat unit.
A second requirement was that the polymer employed should easily
form regular crystals. A polymer which meets both of these
requirements is poly(trans 1,4 butadiene)-PTBD- it has a double
bond in each repeat unit and it was known that in solution at least,
those groups could be quantitatively detected by reaction with
perbenzoic acid to form an epoxide.

    A procedure was developed to prepare mostly monolayer PTBD
crystals of similar size from one solvent, heptane. Reaction of
these crystals in suspension with an epoxidizing agent was then
studied; it was concluded that only a fraction of the total number
of double bonds in each crystal was available for epoxidation.
Crystals were then prepared from other solvents and these were
epoxidized in suspension.

    In order to calculate the average number of monomer units per
fold (counting chain ends at the surface as folds) it is necessary
to know the average crystal thickness. Estimates of this parameter
were first made from shadow casting techniques. Low angle X-ray
results,more recently obtained, on crystals grown from various
solvents show that the number of monomer units per fold depends on
the conditions of crystal growth.

    Having determined the number of monomer units available for
reaction at the surfaces for PTBD crystals, various physical
properties which depend on the amorphous-crystalline ratio were
studied. These included studies of the infrared spectra and heat
of transition of crystal mats in addition to "broad-line" n.m.r.
measurements on crystals wetted with a non-protonated solvent.
These investigations all support the conclusion that in addition
to the surface amorphousness associated with chainfolds and chain
ends there is for some PTBD crystal preparations a significant

amount of interior amorphousness.

CRYSTAL PREPARATION[8]

Two batches of PTBD were employed. One batch (PTBD-K) was found to have a M_n of 8670 ($\pm$10%) for the as received polymer and an M_n of 36,900 for material precipitated from dilute heptane solution. The other batch (PTBD-U) gave a M_v of 11,000.

Crystals were grown from six solvents or solvent-mixtures: heptane, MIBK, toluene, benzene, toluene-ethanol and benzene-ethanol. For the first three solvents the following method was employed. The polymer was dissolved at the minimum temperature for that particular solvent-PTBD system, the solution (0.02% by weight) filtered, the resulting mixture reheated to the original dissolution temperature and the solution placed in a constant temperature bath set low enough to produce a large quantity of crystals of uniform size. After precipitation was complete, the crystals were separated from the growth liquid by filtration. These were washed with pure liquid at room temperature or below. For the epoxidations the crystals were resuspended in benzene and stored at 6°C. In the cases where the crystals were to be used for i.r. or heat of transition measurements a crystal mat was made by filtration. For the n.m.r. measurements the crystals were resuspended in carbon disulfide and stored at 0°C. The crystals prepared from benzene, toluene-ethanol and benzene-ethanol were not prepared by the minimum dissolution technique. (8,9) The dissolution and precipitation temperatures and the % polymer precipitating for the various preparations used are given in Table I.

TABLE I

Characteristics of PTBD Crystals

Sample		Dissol. temp, $^\circ$C	Ppt. temp, $^\circ$C	% Ppt.	$\mathscr{L}$ A	% D.B.
PTBD-K	MIBK	92	73	68		22
	Heptane	78	64	75-78	110	14
	Toluene-ethanol	50	35	-		
	Toluene	50	23	40-76	94	19
	Benzene	52	8	35	103	27
PTBD-U	MIBK	88	55			
	Heptane	69	50	13-25		34
	Toluene-ethanol	50	35	-		
	Benzene-ethanol	42	33	-		

Electron micrographs of PTBD-K and PTBD-U crystals from heptane and PTBD-K crystals from MIBK have been given earlier. (6,8,10)
Electron micrographs of some PTBD-K crystals grown from toluene
and from benzene are given in Figs. 1 and 2, respectively.
Crystals from these two solvents were found to be about one-half
to one-third the size of heptane or MIBK grown crystals. They are
also elongated hexagons whereas those from heptane and MIBK are
regular hexagons. The thicknesses of crystals from heptane,
toluene, and benzene as obtained from low angle X-ray measurements,
however, are about the same, that is 100 ± 10A (see Table I).

EPOXIDATION OF CRYSTALS IN SUSPENSION AND CHAIN FOLDING

The epoxidation of PTBD crystals was carried out using meta-
chloroperbenzoic acid at 0°C. in benzene solution. The metachloro
derivative was used because of its greater stability in solution
in the absence of PTBD; therefore, only a small correction in the
amount of the peroxy acid consumed in the epoxidation was required.
Approximate amounts of mCPBA, large enough to react with all the
available double bonds, were determined in initial runs for each
crystal preparation. In a few cases it was necessary to add
additional mCPBA during a run. Aliquots of the liquid part of the
suspension were removed at various times and the mCPBA remaining
determined by iodometric titration. In all but a few cases the %
double bonds reacting leveled off after an initial rise, then
remaining constant with time. Double bonds reacted vs. time curves
which illustrate this behavior have been shown for PTBD-K crystals
from heptane (three separate runs) (10) and for PTBD-U crystals
from heptane (9) followed for 25 days. Such results for MIBK grown
and toluene grown crystals are given in Figure 3 and 4. The ratio
of moles mCPBA added to moles of double bonds in the particular
sample is noted on the figures. In one run on PTBD-U crystals a
sharp rise was noted (9) in % double bonds reacted after five days
with a leveling off at 50% reacted instead of 34%, as given in
Table I. This exception to the usual observation was attributed to
cracks formed in the crystals during the time they were being
epoxidized. The final % double bonds reacted are given in Table I
for various preparations. Assuming that the PTBD-K and PTBD-U
crystals are not too different in thickness, it can be concluded
from Table I that either there is a difference in the number of
monomer units per fold for these two polymers or the larger value
for the lower molecular weight polymer is due to a larger number
of chain ends present at the crystal surfaces. This latter
explanation is in agreement with results given earlier for polyethylene. (12,13)

From the % double bonds reacted, the crystals thickness and

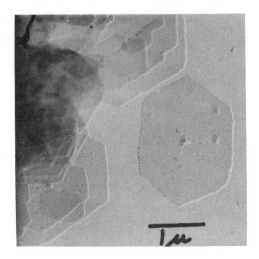

Fig. 1 - PTBD-K crystals from toluene solution - dissolved at
        50°C, ppt at 23°C.

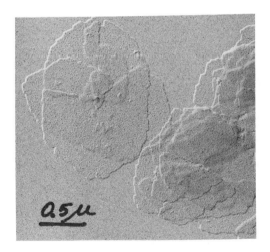

Fig. 2 - PTBD-K crystals from benzene solution - dissolved at 52°C,
        ppt at 8°C.

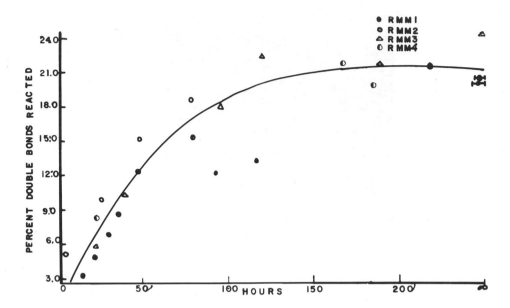

Fig. 3 - % double bonds reacted with mCPBA vs. time for crystals
ppt from MIBK.   [-OOH]/[double bonds]: 1) .27, 2) .41,
3) .26, 4) .23.

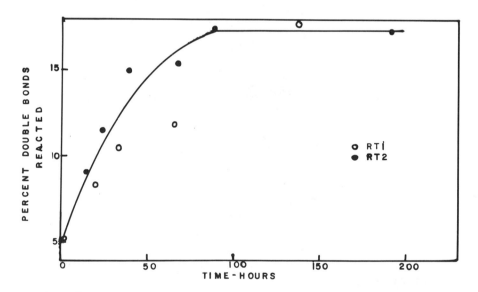

Fig. 4 - % double bond reacted with mCPBA vs. time for crystals
ppt. from toluene. [-OOH]/[double bonds]: 1) .26
2) .23.

the repeat distance of 4.83 Å given by Iwayanagi and coworkers (14) the average number of monomer units per fold, assuming no chain ends in the surface region, is calculated. For PTBD-K crystals from heptane, toluene and benzene values of 3.2, 3.7, and 5.7 respectively, are obtained. If it is assumed that all the chain ends are in the surface regions and that the average length of these is one-half the crystal thickness, then values for the average number of monomer units per fold are calculated to be $2 \frac{1}{2}$, 3 and 5, respectively for heptane, toluene and benzene crystals. It was concluded previously, (8,10) that the tightest fold which could be formed for the low temperature form of PTBD (form I) would contain between $1 \frac{1}{2}$ and 2 monomer units. This suggests that for at least a large number of the folds adjacent reentry is necessary. The experimentally derived values above are average ones - there could be some very loose folds with nonadjacent reentry for any of the preparations studied. However, the presence of very loose folds would necessitate the existence of some tight ones in order to preserve the average values found.

Tatsumi et al (15) postulated from the results of dynamic mechanical studies on PTBD crystals that the fold tightness differed depending on the growth solvent. The results cited above bear out their prediction.

THE AMORPHOUS FRACTION IN PTBD CRYSTALS

The amorphous to crystalline ratio for mats of PTBD crystals grown from various solvents was studied using the i.r. bands at 1335 cm^{-1}, a "regularity" band, and 1350 cm^{-1}, an "amorphous" band. (16) Some spectra were given earlier; some other are shown in Fig. 5. All of the i.r. results obtained are summarized in Table II in terms of I_{1350}/I_{1335}. It is observed that crystals grown at temperatures below the transition temperature e.g. those from toluene, benzene, toluene-ethanol and benzene-ethanol, contain a significantly higher amorphous content than those grown near to or above the transition temperature, e.g. those from heptane and MIBK. In addition, annealing above the crystal-crystal transition temperature, T_{tr}, leads to an increase in the crystallinity for those crystals grown well below T_{tr}. Upon comparison of the i.r. results with those from the epoxidation studies some interesting facts emerge (see Table III). The total amorphous content relative to the crystalline content as seen by i.r., is about the same as the number of monomer units on the crystal surfaces relative to the number in the interior, as given from epoxidation studies, for PTBD-K crystals from heptane and MIBK. For crystals grown from toluene and from benzene the i.r. ratio exceeds the epoxidation ratio by a factor of 3 or 4. These results lead to the following conclusions: 1) the fold surfaces contain

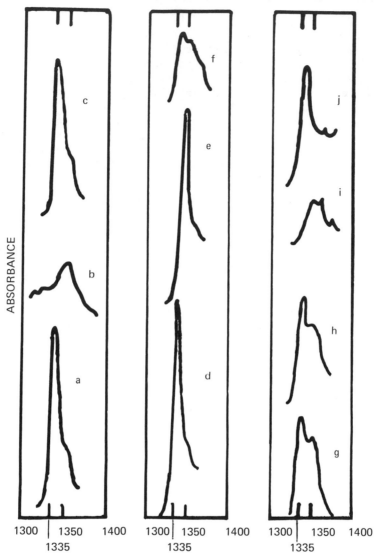

ABSORBANCE

FREQUENCY — cm⁻¹

Fig. 5 - High resolution i.r. absorption peaks at 1335 and 1350 cm⁻¹ for PTBD crystals. a) K crystals from heptane-spectra taken at 25°C. b) same-spectra at 80°C. c) same-spectra at 25°C. d) K crystals from MIBK-spectra taken at 25°C. f) same after melting and quenching to 25°C. g) K crystals from toluene-ethanol-spectra at 25°C. h) U crystals from benzene-ethanol-spectra at 25°C. i) same-spectra at 80°C j) same-spectra at 25°C.

TABLE II[9]

I_{1350}/I_{1335} For PTBD Crystal Mats At 25°

Sample	Solvent	I_{1350}/I_{1335}	
		As prepared	After 80° annealing
PTBD-K	MIBK	0.2	0.2
	Heptane	0.2	0.1
	Toluene-ethanol	1	0.4
	Toluene	1.2	0.6
	Benzene	0.9	0.3
	Benzene cast film	0.3	
	Melt formed	0.5	
PTBD-U	MIBK	0.2	0.3
	Heptane	0.3	0.2
	Toluene-ethanol	1.3	0.2
	Benzene-ethanol	1.2	0.3
	Melt formed	1	

TABLE III[9]

Comparison of Epoxidation and IR

Results for PTBD Crystals

Sample	Solvent	$\alpha/(1 - \alpha)$	I_{1350}/I_{1335}
PTBD-K	MIBK	0.28	0.2
	Heptane	0.16	0.2
	Toluene	0.23	1
	Benzene	0.37	1
PTBD-U	Heptane	0.52	0.3

amorphous material, 2) amorphous regions exist in the interior of crystals grown at temperatures below the transition temperature and 3) interior amorphous regions can be ordered by heating above T_{tr} and subsequently cooling.

The enthalpies of transition and of melting were studied using differential scanning calorimetry, DSC with a Perkin Elmer DSC - 1B on mats of PTBD-K crystals grown from heptane, toluene and benzene. (17) The results are summarized in Table IV. The values

TABLE IV. Calorimetric Data on PTBD-K Crystals Grown From Various Solvents

Solvent Growth Conditions	Transition			Melting			Recrystallization			Transition (melt recryst. sample)		
	ΔH (cal/gm)	ΔS eu	T °C	ΔH (cal/g)	ΔS eu	T °C	ΔH (cal/g)	ΔS eu	T °C	ΔH (cal/g)	ΔS eu	T °C
Heptane (76/63°C)	26±2	.075	73	16.1±.5	.039	139	14.7±.7	.037	120	27±2	.078	71
Toluene (50/23°C)	19±3	.058	55	15.3±.6	.037	136	13.6±.5	.034	123	27±2	.078	71
Benzene (52/8°C)	18±4	.055	55	13.9±.8	.034	135	13±2	.034	120	26±3	.075	74

given are averages for three samples or more from each preparation.

The peaks observed in the DSC scans for the three crystal preparations were regular and sharp, similar in shape to those for the melt-recrystallized samples. It was also found that doubling the scanning rate from 20°C/minute to 40°C/minute did not change the appearance of the peaks for benzene grown crystals. For irradiated toluene-grown crystals, ΔH_t was 21 ± 6 cal/gm; after heating above the melting temperature and cooling, ΔH_t was 20 ± 3 cal/gm. It is seen that this ΔH_t value is close to that for non-irradiated toluene crystals (see Table IV). All of these observations lead to the conclusion that in the crystal-crystal transition region PTBD crystals do not undergo thermal rearrangements large enough to bring about changes in crystallinity. However, it was found that for benzene and toluene grown crystals, both irradiated and non-irradiated, the transition temperature increased as the sample was repeatedly run through the crystal-crystal transition although ΔH_t was not altered. It is concluded that in these samples, which show a low infrared crystallinity ($\sim 50\%$), some internal rearrangements are taking place which, however, do not contribute significantly to ΔH_t.

The heptane-grown single crystals exhibited the greatest reproducibility of results; these also have the highest crystallinities from infrared measurements (9) ($\sim 80\%$). Crystals grown from toluene and benzene did not behave in this regular fashion, as is evident from the error limits given in Table I. The greater degree of amorphousness for these crystals is reflected both in the variability of their thermal behavior and in the large differences in ΔH_t for single crystals versus that for melt recrystallized speciments. Once a single crystal sample is melt recrystallized it loses these difference and the ΔH_t values become more reproducible and independent of crystal growth history.

It is noted from Table IV that the experimental ΔH_t for PTBD crystals from toluene and from benzene are lower than the value for crystals from heptane. Since the crystal structure (8,14) and lamellar thickness, for these various preparations, are about the same, this effect must be caused by differences in the amount of disorder associated with a given preparation. It is found that correction of the ΔH_t values to yield values based on the weight of crystalline material as given by the i.r. results (see Table III) leads to numbers which agree within experimental error for the three preparations. It is concluded that the benzene and toluene prepared crystals contain internal as well as surface disorder as suggested by the i.r. results alone.

Work is currently in progress using a third experimental method that should yield further information about the amorphous

to crystalline ratios in PTBD crystals. This involves the study
of the narrow n.m.r. lines for crystals wetted with a non-
protonated solvent. Only the protons in region available to the
solvent will contribute to the narrow line. For vacuum dried
heptane grown PTBD-K crystals the narrow line disappears at about
250°K. and has a line width in the 250° - 300°K region of 0.5 gauss.
The amorphous to crystalline ratio at 300°K gotten from the ratio
of the narrow-line intensity to the broad-line intensity is 0.1.
For heptane grown crystals where the heptane has been replaced by
carbon disulfide <u>without</u> drying the crystals, the narrow line in
the 209° - 273°K. region is $\leq$ 0.2 gauss in width; in this temperature range the amorphous to crystalline ratio is found to be
< 0.3. At 180°K the narrow line is almost completely absent.
For dry toluene grown crystals the amorphous to crystalline ratio
is 0.3 and the line width at 300°K is 0.3.

The n.m.r. results to date show the following. 1) The total
amorphous content is larger for toluene grown crystals than for
those from heptane. 2) The presence of a liquid, CS_2, greatly
enhances the motion taking place in the amorphous regions of PTBD
crystals.

Further information about PTBD crystals has been gathered
from some preliminary studies on the dissolution of PTBD crystals
by CS_2 at 25°C. When toluene-grown crystals suspended in toluene
are resuspended in CS_2, the crystals dissolve within a few minutes
at 25°C. Heptane crystals given the same treatment are only slowly
attached. After 60 days exposure a number of the crystals appear
to have sustained little if any change. The differences in rate
of solution could possibly be affected by the two-fold difference
in crystal size between toluene - and heptane-grown crystals.
However, this cannot account for the almost instantaneous penetration of the toluene-grown crystals by CS_2; it appears that this is
evidence for a mosaic structure for these crystals, that is, there
are disordered regions which traverse the crystal from one face to
the other or from a face deep into the crystal interior.

SUMMARY

PTBD crystals were grown from dilute solutions of various
solvents above and below the crystal-crystal transformation temperature. A technique to prepare crystals of uniform size was
developed. Epoxidation of the crystals in suspension leads to
reaction of only a fraction of the double bonds in the sample.
Infrared spectra, particularly the bands at 1335 and 1350 cm^{-1},
and differential scanning calorimetry scans through the crystal-
crystal transition region were obtained on dry crystal mats.
Broad-line n.m.r. and some preliminary dissolution measurements

were made on crystals wet with a mon-protonated solvent. From the results of these investigations to date the following conclusions can be given.

1) The number of monomer units on the surfaces of crystals of the same thickness depends on the conditions of growth (solvent and/or growth temperature) and on the molecular weight. The molecular weight dependence is ascribed to the presence of free chain ends. Reentrant folding with some fold looseness is postulated.

2) The i.r. and DSC measurements show that there is an amorphous fraction which exceeds the fraction of chain segments at the surfaces by a significant amount for crystals grown below the crystal-crystal transition temperature. This fraction can be lessened by annealing above the transition-temperature but below the melting point.

3) The n.m.r. and preliminary solvent dissolution studies support the i.r. and DSC measurements. However, the solvent dissolution studies suggest that any amorphous regions which exist in the crystal interior are available to the solvent; this implies the presence of a mosaic structure of amorphous and crystalline regions.

<u>REFERENCES</u>

1. R.K. Sharma and L. Mandelkern, Macromolecules 3 758 (1970).
2. E.W. Fisher, and P. Kloos, J. Polymer Sci. B8, 685 (1970).
3. E.W. Fischer, P. Kloos and G. Lieser, J. Polymer Sci. B7, 845 (1969).
4. A. Peterlin, J. Macromol. Sci (Phys) 3, 19 (1969).
5. D.J. Blundell, A. Keller and T. Connor, J. Polymer Sci A2 5, 991 (1967).
6. A. Keller and Y. Udagawa, J. Polymer Sci. A2 9, 1793 (1971).
7. A. Keller, E. Martuscelli, D.J. Priest and Y. Udagawa, J. Polymer Sci. A2 9, 1807 (1971).
8. J.M. Stellman and A.E. Woodward, J. Polymer Sci. A2 9, 59 (1971).
9. C. Hendrix, D.A. Whiting and A.E. Woodward, Macromolecules 4, 571 (1971).
10. J.M. Stellman and A.E. Woodward, J. Polymer Sci B7, 755 (1969).
11. B. Newman, J.M. Stellman and A.E. Woodward, J. Polymer Sci A2
12. A. Keller and D.J. Priest, J. Macromol. Sci. (Phys) 2, 479 (1968).
13. S. Krimm and M.I. Bank, J. Polymer Sci. A2 7, 1785 (1969).
14. S. Iwayanagi, I. Sakurai, T. Sakurai and T. Seto, J. Macromol. Sci (Phys) B2, 163 (1968).

15. T. Tatsumi, T. Fukushima, K. Imada and M. Takayanagi, J. Macromol. Sci (Phy) B1, 459 (1967).

16. D. Morero, F. Ciampelli and E. Mantica, Adv. Mol. Spectrosc., Proc. Int. Mtg, 4th, 1959 $\underline{2}$, 898 (1962).

17. J.M. Stellman and A.E. Woodward, J. Polymer Sci. A2.

NONLINEAR VISCOELASTIC CONSTITUTIVE BEHAVIOR AS

A STATISTICAL DYNAMIC SYSTEM

W. Chen and C. C. Hsiao

University of Minnesota

Minneapolis, Minnesota 55455

Abstract

In viscoelasticity the governing constitutive equations are usually given in terms of the stress tensor as the functional of the history of deformation or of memory or hereditary functions. Concepts as these are abstract, even fictitious, and lack of physical soundness. However, using statistical irreversible flow processes characterized by joint strain and strain rate fields the statistical distribution of the nonequilibrium states resulted from the deformation of a viscoelastic molecular body by external forces can be determined. Quantities generally referred to as temporal correlation functions play a principal role which permit the derivation and interpretation of nonlinear viscoelastic behavior in the current time domain.

Recently the idea of using statistical considerations for calculating a variety of transport coefficients has been extended to the calculation and derivation of the constitutive equations in linear viscoelasticity. This approach follows a sound physical reasoning and leads into a natural result of the measurable stress tensor as a temporal average of tensors without utilizing history of deformation

or memory function as has been done in the past.
Khazanovich [1] has considered the stress tensor with
small deformation of a linear viscoelastic body.   In
the present paper an attempt is made to obtain a set
of nonlinear viscoelastic constitutive equations
using strain and strain rate distributions rather
than position and velocity distributions.   Under
certain assumptions temporal cross correlation
functions are found to be useful in expressing the
nonlinear terms in current time domain.

On the basis of a simple viscoelastic molecular
model the following attempt is made to derive a
nonlinear constitutive representation of the material
body.   The model is assumed to compose of an arbitrary
division of small individual volumes within each a
sufficiently large number of n elemental material
domains are contained.   The mass center m_i of an
arbitrary representative domain is identifiable by
x_i (i=1,2,3,...n) in a rectangular coordinate system
x_α (α=1,2,3).   Corresponding to these domains in a
one-to-one correspondence there exists a family of
strains ε_i (i=1,2,3,...n) and strain rates $\dot{\varepsilon}_i$ (i=1,2,3..
..n) which describe the motion of the individual
elemental volume body statistically in the phase
space as a result of applied external disturbances.
These disturbances cause deformation of the body
through the creation of a state of thermodynamic
nonequilibrium in the system.   The determination of
the statistical distribution of the nonequilibrium
states becomes the prime task in understanding the
processes of deformation.

For any individual elemental volume the motion
characterized by the distribution of the strains and

strain rates of the mass centers is governed by a state
of thermodynamic equilibrium in that volume.   The state
of equilibrium is describable at a particular time   t
using a statistical equilibrium distribution function
ρ_e of the following exponential form*.

$$\rho_e(\underset{\sim}{\dot{\varepsilon}},\underset{\sim}{\varepsilon},t) = C_o \, e^{-\Lambda(\underset{\sim}{\dot{\varepsilon}},\underset{\sim}{\varepsilon},t)} \tag{1}$$

where C_o is a constant and

$$\Lambda(\underset{\sim}{\dot{\varepsilon}},\underset{\sim}{\varepsilon},t) = H(\underset{\sim}{\dot{\varepsilon}},\underset{\sim}{\varepsilon},t) + \Phi(\underset{\sim}{\dot{\varepsilon}},\underset{\sim}{\varepsilon},t) \tag{2}$$

Here   H   is associated with the dimensionless energy
of the system in the following form

$$H(\underset{\sim}{\dot{\varepsilon}},\underset{\sim}{\varepsilon},t) = \sum_i (\lambda_1 \dot{\varepsilon}_{\sim i}^2 + \lambda_2 \varepsilon_{\sim i}^2) \tag{3}$$

and Φ is associated with the dimensionless external
energy resulted from boundary forces outside of the
elemental volume.   These dimensionless energy quantities can be easily seen as the ratio of the combination of kinetic energy, strain energy, and work over
the usual thermal energy RT.   For simplicity without
losing generalities the absolute temperature   T   is
assumed to be constant during the process.   Since   R
is a universal constant the thermal energy is easily
incorporated into the energy quantities.

Observe that the distribution function $\rho_e(\underset{\sim}{\dot{\varepsilon}},\underset{\sim}{\varepsilon},t)$
represents the equilibrium state of the material system
within an elemental volume at time   t.   The changes in
the distributions are continually restoring equilibrium
and erase the effect of disturbances.   At a later time
the distribution function may be modified by introduc-
ing an additional function $\Delta(\underset{\sim}{\dot{\varepsilon}},\underset{\sim}{\varepsilon},t)$ to account for
the nonequilibrium characteristics as follows:

---

* Readers not well aquainted with statistical dynamic
considerations may refer to Appendix A for an approach
in obtaining the exponential distribution function.

$$\rho = C_o e^{-\Lambda + \Delta} = \rho_e e^{\Delta} \tag{4}$$

This approach is concerned with the direct determination of the change in time of the state variables and
the transport coefficients will be expressed in terms
of correlation functions of equilibrium fluctuations
of the appropriate dynamical quantities. The explicit
introduction of nonequilibrium distribution functions
is not necessary. As a result the nonequilibrium
stress field may be considered as the equilibrium
stress field $\underset{\sim}{\sigma}_e$ modified by the nonequilibrium distribution function ρ in the following simple manner
where Δ is in general small for rapid convergence
of the series. However the expansion need not require
Δ to be small.

$$\underset{\sim}{\sigma} = \underset{\sim}{\sigma}_e \rho = \underset{\sim}{\sigma}_e \rho_e e^{\Delta}$$
$$= \underset{\sim}{\sigma}_e \rho_e (1 + \Delta + \frac{1}{2!}\Delta^2 + \frac{1}{3!}\Delta^3 + \ldots) \tag{5}$$

So far in the elemental volume there is a sufficiently large number of representative material elements yet the volume is small enough so that the
distribution function may be considered to vary in a
continuous manner. This satisfies the invariance
conditions in phase-space and according to the
Liouville's theorem the distribution function of
representative strains and strain rates corresponding
to the motion of a system of material elements remains
constant during the motion. That is (4) must obey
the following condition:

$$\frac{d\rho}{dt} \equiv 0 = -C_o e^{-\Lambda+\Delta} \frac{d(\Lambda-\Delta)}{dt} \tag{6}$$

As a result with proper zero initial condition for Δ:

$$\Delta = \int_o^t \frac{d\Lambda}{dt} dt \tag{7}$$

Now returning to (2)

$$\frac{d\Lambda}{dt} = \frac{\partial H}{\partial t} + \sum_i \left(\frac{\partial H}{\partial \varepsilon_i} \frac{d\varepsilon_i}{dt} + \frac{\partial H}{\partial \dot{\varepsilon}_i}\frac{d\dot{\varepsilon}_i}{dt}\right)$$

$$+ \frac{\partial \Phi}{\partial t} + \sum_i \left(\frac{\partial \Phi}{\partial \varepsilon_i} \frac{d\varepsilon_i}{dt} + \frac{\partial \Phi}{\partial \dot{\varepsilon}_i}\frac{d\dot{\varepsilon}_i}{dt}\right) \tag{8}$$

Application of the variational principle to (3) one obtains the canonical properties of $H(\varepsilon_i, \dot{\varepsilon}_i, t)$.

$$\frac{d\varepsilon_i}{dt} = \frac{\partial H}{2\lambda_1 \partial \dot{\varepsilon}_i} \tag{9}$$

$$\frac{d\dot{\varepsilon}_i}{dt} = -\frac{\partial H}{2\lambda_1 \partial \varepsilon_i} \tag{10}$$

then (8) reduces to

$$\frac{d\Lambda}{dt} = \frac{\partial H}{\partial t} + \frac{\partial \Phi}{\partial t} + \sum_i \left(\frac{\partial \Phi}{\partial \varepsilon_i} \dot{\varepsilon}_i + \frac{\partial \Phi}{\partial \dot{\varepsilon}_i} \ddot{\varepsilon}_i\right) \tag{11}$$

Here for local equilibrium H need not be an explicit function of time implying $\partial H/\partial t = 0$, Φ is dependent upon time only through the dependence of the deformation on time. Thus

$$\frac{\partial \Phi}{\partial t} = 0 \tag{12}$$

The last term in (11) represents a combination of linear dissipative force and an acceleration term which can be modified to give a volume integral with the aid of (10). Combine this with the remaining summation term in (11) by a volume integral, one gets

$$\frac{d\Lambda}{dt} = \int_{dv} \left(\frac{\partial \Phi}{\partial \varepsilon} \dot{\varepsilon} + D\dot{\varepsilon} \frac{1}{2\lambda_1} \frac{\partial H}{\partial \varepsilon}\right) dv \equiv p\dot{\varepsilon}dv \tag{13}$$

with $\dot{\varepsilon}$ regarded as a statistical average over the small volume dv so that (13) is obtainable in its form where D is the dissipative coefficient, a constant, and p is the viscous stress tensor in the equilibrium state. Here the strain rate field is considered

invariant throughout the small volume with respect
to which the integration is carried out.

Because of macroscopic inhomogeneities in describing the stress field in the vicinity of the representative elemental volume it is necessary to obtain
the stress tensor as the statistical mean of a tensor.
Employing an extended virial theorem that the time
average of a sort of energy quantity of a system is
equal to its virial in a general form as shown in
Appendix B

$$< \sum_i m_i \dot{x}_i(t)\dot{x}_i(\tau) > = -< \sum_i x_i(t) F_i(\tau) > \qquad (14)$$

where t is the current time, τ is another time and
the brackets designate the average over a sufficient
length of time provided the quantities involved
including velocities are finite and bounded. Here F_i
represents the forces including forces of constraint
associated with the ith material domain in the
elemental volume. The contributions to this force
are both from within and without the boundary of the
volume. As mentioned earlier the equilibrium stress
field must be obtained as the statistical mean of a
tensor. In the present case it is the time average
of a stress tensor π whose value is given in Appendix C.

$$\sigma_e = <\pi> \qquad (15)$$

Now evaluating (5) by averaging with the aid of (7)
over a sufficient period of time, one obtains the
nonequilibrium stress tensor in terms of temporal
correlation functions

$$\sigma(t) = \rho_e <\pi(t)> + \int_0^t \rho_e <\pi(t) \frac{d\Lambda}{dt}(\tau)> d\tau$$

$$+ \int_0^t \int_0^t \rho_e << \frac{1}{2!} \pi(t) \frac{d\Lambda}{dt}(\tau_1) > \frac{d\Lambda}{dt}(\tau_2) > d\tau_1 d\tau_2$$

$$+ \int_0^t \int_0^t \int_0^t \rho_e <<< \frac{1}{3!} \pi(t) \frac{d\Lambda}{dt}(\tau_1) > \frac{d\Lambda}{dt}(\tau_2) > \frac{d\Lambda}{dt}(\tau_3) > \cdot$$

$$d\tau_1 d\tau_2 d\tau_3 \ + \ \dots \tag{16}$$

where

$$< \pi(t) \frac{d\Lambda}{dt}(\tau) > \ = \ \frac{1}{t-\tau} \int_0^{t-\tau} \pi(t) \frac{d\Lambda}{dt}(\tau) \, dt \tag{17}$$

is the temporal correlation [2] of two functions $\pi(t)$
of time t and $d\Lambda(\tau)/dt$ of time τ. As indicated
before that the strain rate field contained in $d\Lambda/dt$
is considered invariant throughout the small volume,
if a relatively slow and uniform motion is assumed
so that the strain rate tensor is also constant with
respect to the temporal correlation of the stress
fields, then, using (13)

$$< \pi(t) \, p(\tau) \, \dot{\varepsilon}(\tau) > \ = \ < \pi(t) \, p(\tau) > \dot{\varepsilon}(\tau) \tag{18}$$

and thus (16) can be reduced to

$$g(L) \ - \ \mu_e <_{\|}(t) > \ + \ \int_0^t \rho_e < \pi(t) \, p(\tau) > \dot{\varepsilon}(\tau) \, d\tau$$

$$+ \int_0^t \int_0^t \rho_e << \frac{1}{2!} \pi(t) \, p(\tau_1) > p(\tau_2) > \dot{\varepsilon}(\tau_1) \, \dot{\varepsilon}(\tau_2) \, d\tau_1 d\tau_2 d\tau_3$$

$$+ \int_0^t \int_0^t \int_0^t \rho_e <<< \frac{1}{3!} \pi(t) \, p(\tau_1) > p(\tau_2) > p(\tau_3) > \cdot$$

$$\dot{\varepsilon}(\tau_1) \, \dot{\varepsilon}(\tau_2) \, \dot{\varepsilon}(\tau_3) \, d\tau_1 d\tau_2 d\tau_3$$

$$+ \ \dots \tag{19}$$

The temporal correlations [2] may be reduced as
follows:

$$\rho_e < \pi(t) \, p(\tau) > \ = \rho_e \frac{1}{t-\tau} \int_0^{t-\tau} \pi(t) \, p(\tau) \, d\tau$$

$$= \frac{\rho e}{s} \int_0^s \underset{\sim}{\pi}(\tau+s)\underset{\sim}{p}(\tau)d\tau = \underset{\sim}{C}(s) = \underset{\sim}{C}(t-\tau) \qquad (20)$$

where $\underset{\sim}{C}(t-\tau)$ is a tensor quantity of the fourth rank, and for the cross correlation operations:

$$\frac{1}{2}\rho_e <\underset{\sim}{C}(t,s_1)\underset{\sim}{p}(\tau_2)> = \frac{\rho e}{2!(t-\tau_2)} \int_0^{t-\tau_2} \underset{\sim}{C}(t,s_1)\underset{\sim}{p}(\tau_2)d\tau_2$$

$$= \frac{\rho e}{2s_2} \int_0^{s_2} \underset{\sim}{C}(\tau_2+s_2,s_1)\underset{\sim}{p}(\tau_2)d\tau_2 \qquad (21)$$

$$= \underset{\sim}{C}(s_1,s_2) = \underset{\sim}{C}(t-\tau_1,t-\tau_2)$$

where $\underset{\sim}{C}(t-\tau_1,t-\tau_2)$ is a tensor of the sixth rank. By substitution finally (19) yields the nonlinear constitutive equation

$$\underset{\sim}{\sigma}(t) = \underset{\sim}{C}(0)\underset{\sim}{\varepsilon}(t) + \underset{\sim}{D}(0)\dot{\underset{\sim}{\varepsilon}}(t) + \int_0^t \underset{\sim}{C}(t-\tau)\dot{\underset{\sim}{\varepsilon}}(\tau)d\tau$$

$$+ \int_0^t \int_0^t \underset{\sim}{C}(t-\tau_1,t-\tau_2)\dot{\underset{\sim}{\varepsilon}}(\tau_1)\dot{\underset{\sim}{\varepsilon}}(\tau_2)d\tau_1 d\tau_2$$

$$+ \int_0^t \int_0^t \int_0^t \underset{\sim}{C}(t-\tau_1,t-\tau_2,t-\tau_3)\dot{\underset{\sim}{\varepsilon}}(\tau_1)\dot{\underset{\sim}{\varepsilon}}(\tau_2)\dot{\underset{\sim}{\varepsilon}}(\tau_3) \cdot$$

$$d\tau_1 d\tau_2 d\tau_3 + \dots \qquad (22)$$

where $\underset{\sim}{C}(0)\underset{\sim}{\varepsilon}(t)$ represents linear elastic behavior and $\underset{\sim}{D}(0)\dot{\underset{\sim}{\varepsilon}}(t)$ represents linear fluid behavior and the other terms are additional linear and nonlinear contributions. In component form

$$\sigma_{\alpha\beta}(t) = C_{\alpha\beta\gamma\delta}(0)\varepsilon_{\gamma\delta}(t) + D_{\alpha\beta\gamma\delta}(0)\dot{\varepsilon}_{\gamma\delta}(t)$$

$$+ \int_0^t C_{\alpha\beta\gamma\delta}(t-\tau)\dot{\varepsilon}_{\gamma\delta}(\tau)d\tau$$

$$+ \int_0^t \int_0^t C_{\alpha\beta\gamma\delta\xi\eta}(t-\tau_1,t-\tau_2)\dot{\varepsilon}_{\gamma\delta}(\tau_1)\dot{\varepsilon}_{\xi\eta}(\tau_2)d\tau_1 d\tau_2$$

$$+ \int_0^t \int_0^t \int_0^t C_{\alpha\beta\gamma\delta\xi\eta\lambda\mu}(t-\tau_1,t-\tau_2,t-\tau_3) \cdot$$

$$\dot{\varepsilon}_{\gamma\delta}(\tau_1)\dot{\varepsilon}_{\xi\eta}(\tau_2)\dot{\varepsilon}_{\lambda\mu}(\tau_3) \ d\tau_1 d\tau_2 d\tau_3 + \dots \qquad (23)$$

represents small nonlinear deformation of a viscoelastic medium.

Appendix A

Consider an infinitesimal finite volume containing
a large number of material elements in a viscoelastic
material body. Associated with each material element
the position of its mass center is identified by
x_i (i=1,2,...n) and its velocity $\dot{x}_i$ which characterize
its motion. Corresponding to a particular motion
there exists a family of strain rates $\dot{\varepsilon}_i$ and strains
ε_i which determine its equilibrium state. Through
mapping the probability of locating the end points
of the family of strain rates between the ellipsoidal
shell $\dot{\varepsilon}$ and $\dot{\varepsilon}+d\dot{\varepsilon}$ in a rectangular coordinate system
can be found as follows:

$$\rho_e(\dot{\varepsilon}) = f_1(\dot{\varepsilon}_{x_1}) f_2(\dot{\varepsilon}_{x_2}) f_3(\dot{\varepsilon}_{x_3}) \qquad \text{(A1)}$$

Here the anisotropic nature of the strain rate vectors
is taken into consideration and the principal directions
of the strain rate ellipsoid is put coinciding with
the three coordinate axes. In this expression $\rho_e(\dot{\varepsilon})$
is the equilibrium distribution function and f's are
the functions to be determined. The variation of ρ_e is

$$\delta\rho_e = 0 = \frac{\partial\rho}{\partial\dot{\varepsilon}_{x_1}}\delta\dot{\varepsilon}_{x_1} + \frac{\partial\rho}{\partial\dot{\varepsilon}_{x_2}}\delta\dot{\varepsilon}_{x_2} + \frac{\partial\rho}{\partial\dot{\varepsilon}_{x_3}}\delta\dot{\varepsilon}_{x_3} \qquad \text{(A2)}$$

Combine (A1) and (A2) and let $f' \equiv \partial f(x)/\partial x$ then

$$0 = \frac{f_1'(\dot{\varepsilon}_{x_1})}{f_1(\dot{\varepsilon}_{x_1})}\delta\dot{\varepsilon}_{x_1} + \frac{f_2'(\dot{\varepsilon}_{x_2})}{f_2(\dot{\varepsilon}_{x_2})}\delta\dot{\varepsilon}_{x_2} + \frac{f_3'(\dot{\varepsilon}_{x_3})}{f_3(\dot{\varepsilon}_{x_3})}\delta\dot{\varepsilon}_{x_3} \qquad \text{(A3)}$$

In order that the strain rates are to stay in an
ellipsoidal shell region between $\dot{\varepsilon}$ and $\dot{\varepsilon} + d\dot{\varepsilon}$ the

strain rate vectors must satisfy the relation in the
rate-space that

$$\frac{\dot{\varepsilon}^2_{x_1}}{a_1^2} + \frac{\dot{\varepsilon}^2_{x_2}}{b_1^2} + \frac{\dot{\varepsilon}^2_{x_3}}{c_1^2} = \text{constant} \tag{A4}$$

where $a_1, b_1, c_1,$ are the axes of the strain rate
ellipsoid. The variational condition gives

$$\frac{\dot{\varepsilon}_{x_1}\delta\dot{\varepsilon}_{x_1}}{a_1^2} + \frac{\dot{\varepsilon}_{x_2}\delta\dot{\varepsilon}_{x_2}}{b_1^2} + \frac{\dot{\varepsilon}_{x_3}\delta\dot{\varepsilon}_{x_3}}{c_1^2} = 0 \tag{A5}$$

Comparing (A5) with (A3)

$$\frac{f_1'(\dot{\varepsilon}_{x_1})}{f_1(\dot{\varepsilon}_{x_1})} = -\lambda_1\dot{\varepsilon}_{x_1}/2a_1^2 \tag{A6}$$

where λ_1 is a multiplier. Integration of (A6) yields

$$f_1(\dot{\varepsilon}_{x_1}) = \zeta_1 e^{-\lambda_1\dot{\varepsilon}_{x_1}/2a_1^2} \tag{A7}$$

where ζ_1 is a constant and similarly

$$f_2(\dot{\varepsilon}_{x_2}) = \zeta_1 e^{-\lambda_1\dot{\varepsilon}_{x_2}/2b_1^2} \tag{A8}$$

$$f_3(\dot{\varepsilon}_{x_3}) = \zeta_1 e^{-\lambda_1\dot{\varepsilon}_{x_3}/2c_1^2} \tag{A9}$$

Consequently

$$\rho_e(\underset{\sim}{\dot{\varepsilon}}) = \zeta_1 e^{-\lambda_1(\frac{\dot{\varepsilon}_{x_1}^2}{a_1^2} + \frac{\dot{\varepsilon}_{x_2}^2}{b_1^2} + \frac{\dot{\varepsilon}_{x_3}^2}{c_1^2})/2} \tag{A10}$$

On a somewhat similar basis the distribution
function of the strains is also of the exponential
form in the strain space

$$\rho_e(\underset{\sim}{\varepsilon}) = \zeta_2 e^{-\lambda_2(\frac{\varepsilon_{x1}^2}{a_2^2} + \frac{\varepsilon_{x2}^2}{b_2^2} + \frac{\varepsilon_{x3}^2}{c_2^2})/2} \qquad (A11)$$

where a_2, b_2, c_2 are axes of the strain ellipsoid
and ζ_2 is a constant and λ_2 is a multiplier.

Considering the natural state of a viscoelastic
material body without any anisotropy the principle of
frame indifferent must be satisfied. Like in the
statistical treatment of gas molecules the positions
and momenta form a phase-space without any directional
preference, a viscoelastic body can be visualized as
an orientationless system in a phase-space characterized
by strains and strain rates. Thus the equilibrium
distribution is spherical instead of ellipsoidal.
Imposing the condition that

$$a_1 = b_1 = c_1 = c_2 = b_2 = a_2 = 1/\sqrt{2}$$

for spherical symmetry the joint equilibrium distribution as a function of strains and strain rates
becomes

$$\rho_e(\underset{\sim}{\dot{\varepsilon}}, \underset{\sim}{\varepsilon}) = \rho_e(\underset{\sim}{\dot{\varepsilon}})\rho_e(\underset{\sim}{\varepsilon})$$

$$= \rho_e(\underset{\sim}{\dot{\varepsilon}_1}) \cdots \rho_e(\underset{\sim}{\dot{\varepsilon}_n})\rho_e(\underset{\sim}{\varepsilon_1}) \cdots \rho_e(\underset{\sim}{\varepsilon_n})$$

$$= \zeta_1^n \zeta_2^n e^{-(\Sigma_i \lambda_1 \dot{\varepsilon}_{\sim i}^2 + \Sigma_i \lambda_2 \varepsilon_{\sim i}^2)}$$

$$= C_o e^{-\Sigma_i (\lambda_1 \dot{\varepsilon}_{\sim i}^2 + \lambda_2 \varepsilon_{\sim i}^2)} = C_o e^{-H} \qquad (A12)$$

where $C_o \equiv \zeta_1^n \zeta_2^n$ is the constant as given in the text.
Here the external energy is not included in this
expression. If an external energy term Φ is taken
into consideration then the distribution function ρ_e
would be a function of $H + \Phi$.

Appendix B

Consider a general material system composed of

domains represented by their mass centers m_i (i=1,2,..n) in an elemental volume dV and their corresponding locations x_i prescribed with applied forces F_i including any forces of constraint. The fundamental equations of motion in component form in a rectangular coordinate system x_α (α=1,2 or 3) are representable as follows:

$$m_i \ddot{x}_{i\alpha} = F_{i\alpha} \quad \begin{array}{l} (i = 1,2,\ldots,n) \\ (\alpha = 1,2,\text{or } 3) \end{array} \tag{B1}$$

Of interest is a tensor quantity

$$G_{\alpha\beta} = \sum_i m_i x_{i\alpha} \dot{x}_{i\beta} \quad (\alpha,\beta = 1,2,3) \tag{B2}$$

where the dot designates the time derivative. The total time rate of change of $G_{\alpha\beta}$ is

$$\dot{G}_{\alpha\beta} = \sum_i m_i \dot{x}_{i\alpha} \dot{x}_{i\beta} + \sum_i m_i x_{i\alpha} \ddot{x}_{i\beta} \tag{B3}$$

The last term in (B3) is simply related to force components in the following manner:

$$\sum_i m_i x_{i\alpha} \ddot{x}_{i\beta} = \sum_i x_{i\alpha} F_{i\beta} \tag{B4}$$

which allows (B3) to become

$$\dot{G}_{\alpha\beta} = \sum_i m_i \dot{x}_{i\alpha} \dot{x}_{i\beta} + \sum_i x_{i\alpha} F_{i\beta} \tag{B5}$$

The temporal average of (B5) over a time interval τ is obtained by integrating both sides with respect to t from 0 to τ and dividing by τ :

$$<\dot{G}_{\alpha\beta}> = \frac{1}{\tau} \int_0^\tau \dot{G}_{\alpha\beta} dt = <\sum_i m_i \dot{x}_{i\alpha} \dot{x}_{i\beta}> + <\sum_i x_{i\alpha} F_{i\beta}> \tag{B6}$$

The temporal average of $<\dot{G}_{\alpha\beta}>$ means that

$$<\dot{G}_{\alpha\beta}> = \frac{1}{\tau} [G_{\alpha\beta}(\tau) - G_{\alpha\beta}(0)] \tag{B7}$$

If the motion is periodic, i.e., all coordinates repeat after a certain time and if τ is chosen to be the period, then (B7) vanishes. A similar conclusion can be reached even if the motion is not periodic,

provided the coordinates and the velocities for all
mass domains remain finite so that there is an upper
bound to $G_{\alpha\beta}$. By choosing τ sufficiently large (B7)
can be made as small as desired. In both cases it
then follows that from (B6)

$$\langle \sum_i m_i \dot{x}_{i\alpha} \dot{x}_{i\beta} \rangle = -\langle \sum_i x_{i\alpha} F_{i\beta} \rangle \tag{B8}$$

Or in matrix form with time dependency:

$$\langle \sum_i m_i \dot{\underset{\sim}{x}}_i(t) \dot{\underset{\sim}{x}}_i(\tau) \rangle = -\langle \sum_i \underset{\sim}{x}_i(t) \underset{\sim}{F}_i(\tau) \rangle \tag{B9}$$

Appendix C

In order to see how the stress tensor is related
to the prescribed forces on the mass domains in an
elemental volume the force $\underset{\sim}{F}_i$ as employed in Appendix
B is divided into two parts. The external force
acting on the ith mass domain from outside of the system
$\underset{\sim}{f}_i$ is one part, and the other is the internal force $\underset{\sim}{f}_{ij}$
due to the jth mass domain. That is

$$\underset{\sim}{F}_i = \underset{\sim}{f}_i + \sum_{j \neq i} \underset{\sim}{f}_{ij} \tag{C1}$$

Substituting into (B9), the work function splits into
two terms

$$\langle \sum_i m_i \dot{\underset{\sim}{x}}_i \dot{\underset{\sim}{x}}_i \rangle = -\langle \sum_i \underset{\sim}{x}_i \underset{\sim}{f}_i \rangle - \langle \sum_i \underset{\sim}{x}_i \sum_{j \neq i} \underset{\sim}{f}_{ij} \rangle \tag{C2}$$

The last energy term of (C2) is associated essentially
with intermolecular forces of the system in the
elemental volume which may also be represented as
follows:

$$\langle \sum_i \underset{\sim}{x}_i \sum_{j \neq i} \underset{\sim}{f}_{ij} \rangle = \frac{1}{2} \langle \sum_{ij} (\underset{\sim}{x}_i - \underset{\sim}{x}_j) \underset{\sim}{f}_{ij} \rangle \tag{C3}$$

This internal contribution together with the external
energy contribution represented by the first term of
(C2) are responsible for the dynamical stress field,

i.e. from (C2) and (C3)

$$<\sum_i m_i \dot{\underset{\sim}{x}}_i \dot{\underset{\sim}{x}}_i> + \frac{1}{2}<\sum_{ij} (\underset{\sim}{x}_i - \underset{\sim}{x}_j) \underset{\sim}{f}_{ij}> = -<\sum_i \underset{\sim}{x}_i \underset{\sim}{f}_i> \qquad \text{(C4)}$$

which represents the total external contribution leading to a stress tensor $\underset{\sim}{\pi}$.

$$-<\sum_i \underset{\sim}{x}_i \underset{\sim}{f}_i> = <\underset{\sim}{\pi}>dV \qquad \text{(C5)}$$

Observe that in component form (C5) becomes

$$-<\sum_i x_{i\alpha} f_{i\beta}> = <\oint x_\alpha \pi_{\gamma\beta} n_\gamma dA> \qquad \text{(C6)}$$

where the surface integral is taken with respect to the whole surface of the elemental volume dV and $\pi_{\beta\gamma}$ are components of the pressure tensor on an area dA with n_γ as its outward normal. Assuming proper continuous conditions and invoking the well known divergence theorem the surface integral in (C6) may be represented as follows:

$$\oint x_\alpha \pi_{\gamma\beta} n_\gamma dA = \int_{dV} \frac{\partial x_\alpha \pi_{\gamma\beta}}{\partial x_\gamma} dV$$

$$= \int_{dV} \delta_{\alpha\gamma} \pi_{\gamma\beta} dV + \int_{dV} x_\alpha \pi_{\beta\gamma,\gamma} dV \qquad \text{(C7)}$$

Since the elemental volume is in equilibrium which implies that $\pi_{\beta\gamma,\gamma} = 0$ if body force and inertia force are neglected. Then (C6) reduces to

$$\int_{dV} \pi_{\alpha\beta} dV = \pi_{\alpha\beta} dV = \underset{\sim}{\pi} dV \qquad \text{(C8)}$$

Finally using (C4) and (C5) the equilibrium stress tensor may be represented as follows:

$$\underset{\sim}{\sigma}_e = <\underset{\sim}{\pi}> = \frac{1}{dV}[<\sum_i m_i \dot{\underset{\sim}{x}}_i \dot{\underset{\sim}{x}}_i> + \frac{1}{2}<\sum_{ij} (\underset{\sim}{x}_i - \underset{\sim}{x}_j) \underset{\sim}{f}_{ij}>] \qquad \text{(C9)}$$

References

[1] T. N. Khazanovich, "Derivation of the Equations of Linear Viscoelasticity" PMM Vol. 28, No. 6,

p. 1123, 1964

[2] Y. W. Lee, Statistical Theory of Communication, John Wiley, 1960.

Acknowledgement

The authors wish to express their appreciation for having the opportunity of discussing many points with P. Brunn.

THE INFLUENCE OF PRESSURE ON THE MECHANICAL BEHAVIOR OF

POLYCHLOROTRIFLUOROETHYLENE

[*]A. A. Silano and [+]K. D. Pae

[*]Newark State College, Union, N.J.
[+]Rutgers Univ., New Brunswick, N.J.

ABSTRACT

This investigation examines semi-crystalline poly-chlorotrifluoroethylene (PCTFE) and compares its mechanical behavior under the influence of pressure with that of high crystalline polytrifluoroethylene (PTFE). Both polymers show marked increases in elastic modulus, yield stress, and fracture strength with increases in pressure but variations are evident. For example, differences in tensile stress-strain curves, yield stress, fracture strain, and mode of cold-drawing were observed. Similarities do exist in compressive elastic modulus - pressure relationships. In fact, PCTFE and PTFE exhibit linear compressive modulus-pressure, curves, with discontinuities, which are believed influenced by crystalline phase transitions (PTFE) and low temperature dynamic mechanical relaxations (PCTFE and PTFE). A fine structure model (inter-lamellar amorphous model) for a quick-quenched crystallized polymer is adopted to describe the observed results.

INTRODUCTION

The purpose of this paper is to evaluate the influence of hydrostatic pressure on the mechanical behavior of polychlorotrifluoroethylene (PCTFE) and to compare its mechanical behavior with that of polytrifluoroethylene (PTFE). Weir studied the compressibility of several polymers including PCTFE and PTFE (1). He observed a high-pressure crystalline phase transition (Phase II to Phase III) in PTFE at about 80,000 psi. Warfield examined the compressibility of PCTFE at constant temperature (2). His measurements of the coefficient

131

of compressibility increased from 5.1 x 10^{-5}/atm at
atmospheric pressure and 21°C to 9 x 10^{-5}/atm at 182°C. Pae
et al investigated the mechanical behavior of PTFE up to
100,000 psi (3). They observed a discontinuity in the elastic
modulus – pressure curve near 80,000 psi which was attributed
to the crystalline phase transition. In general, most authors
agree that for most polymers tested there is a corresponding
increase in the elastic modulus, yield stress, and fracture
strength with increases in hydrostatic pressure (4, 5, 6, 7).
It should be noted that both increases and decreases in
ductility have also been observed, depending on the structure
of polymers. However, disagreement exists as to the effect of
the fluid medium on elongation properties of polymers (8, 9).

EXPERIMENTAL APPARATUS AND PROCEDURES

     The apparatus used in these experiments has been described
elsewhere (10). The high pressure medium used in all tests
was a mixture of eight parts of kerosene to one part Shell
Diala AX oil. Pretest soaking of several samples in this
mixture under pressure and a subsequent weighing of each
sample revealed no absorption of the fluid.

     Test samples were machined from commercial grade PCTFE
(KELF 81) extruded rods (11). The average mass density of all
test samples was 2.1299 g/cm^3 which corresponds to approximately
45% crystallinity (12). Compression samples were 1 in. long
cylinders having a diameter of 0.629 in. Tensile samples were
cut to an overall length of 2 in. and threaded at both ends.
The center portion was reduced to a diameter of 0.188 in.
producing a gage length of 1 in. with 0.250 in. radius fillets
at each end. Two samples were tested at each pressure level
to assure that replication was observed.

EXPERIMENTAL RESULTS

     The nominal compressive stress–strain curves obtained at
various pressures up to 100,000 psi. are shown in Fig. 1. It is
evident that Young's modulus increases with pressure as shown
in Fig. 2. The elastic modulus is a linear function of the
pressure with slope 3.03 from atmospheric pressure to 60,000 psi.
and with another slope 2.08 from 60,000 psi. to 100,000 psi.
More will be said about the discontinuity at 60,000 psi. in the
discussion of these data. Average values for the compressive
modulus are tabulated in Table 1. The value of the compressive
modulus at atmospheric pressure is 1.69 x 10^5 psi. as compared
to 1.70 x 10^5 psi. reported by the manufacturer.

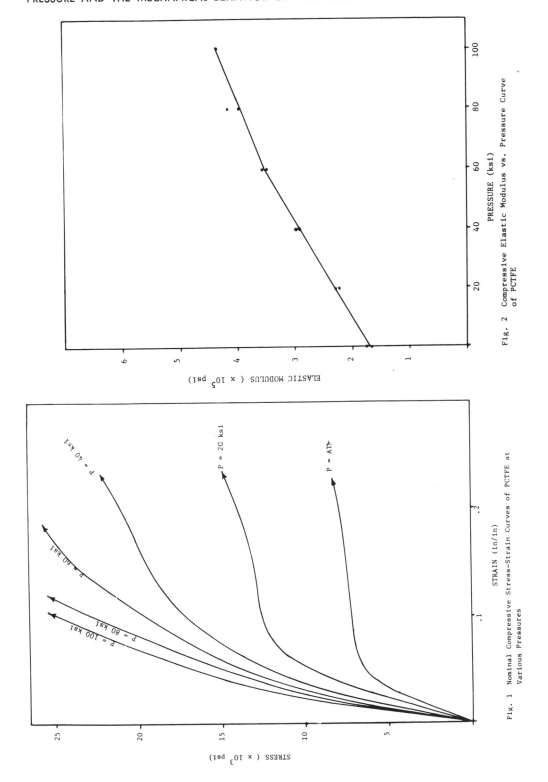

Fig. 2   Compressive Elastic Modulus vs. Pressure Curve of PCTFE

Fig. 1   Nominal Compressive Stress-Strain Curves of PCTFE at Various Pressures

TABLE 1

Average Values of Elastic Modulus at Various Pressures

Pressure (Psi.)	Compressive Modulus (x 10^5 psi.)
Atmospheric	1.69
20,000	2.35
40,000	2.94
60,000	3.53
70,000	3.73
80,000	4.05
90,000	4.21
100,000	4.37

TABLE 2

Average Values of Tensile Peak Yield Stress and
Strain at Various Pressures

Pressure (psi.)	Peak Yield Stress (psi.)	Fracture Strain (in/in)
Atmospheric	5460	.94
20,000	8980	.39
40,000	12,885	.13
60,000	--	.06
80,000	--	.06
100,000	--	.05

Nominal tensile stress–strain curves at various pressures are shown in Fig. 3. No attempt is made to compute the tensile modulus because deformation is measured along the entire sample, including the threaded portion, gage length, and fillets. It is obvious from Fig. 3 that a relative increase in the tensile modulus with pressure occurs.

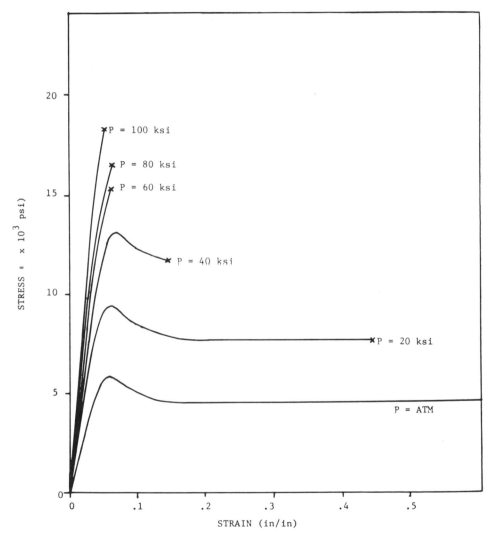

Fig. 3   Nominal Tensile Stress–Strain Curves of PCTFE at Various Pressures

A linear increase in tensile peak yield stress with pressure is evident in Fig. 4, while the strain to fracture decreases rapidly as seen in Fig. 5. Average values for the peak yield stress are tabulated along with fracture strains in Table 2. The presence of a peak stress on nominal stress-strain curves for polymers is usually associated with neck formation on the gage length of the specimen. At the upper yield stress, the height of the peak, the neck develops, and at the lower yield stress, propagation of polymer material in the shoulders of the neck or cold-drawing begins.

Another phenomenon observed in the tensile specimens of PCTFE was crazing or stress whitening. Kambour has emphasized that crazes are not cracks (13). The interesting feature of crazing observed in PCTFE is that it appears to initiate at the interior of the cross section and then propagates radially to the surface of the specimen while also moving longitudinally along the gage length. The fact that the application of pressure suppresses craze formation has been reported (14). It has been observed that craze formation originates in the necked region and propagated in the direction of the applied stress. To evaluate the effect of the pressure medium on stress-strain behavior and crazing in PCTFE, identical tests were conducted in a pentane medium, with comparable results.

DISCUSSION OF RESULTS

The PCTFE used in this study is considered a medium crystalline polymer (~45%) (11). It differs structurally from PTFE in having one chlorine atom in place of a fluorine atom in the monomer. The high dipole moment of the chlorine atom permits easy moldability and increased toughness. X-ray data show that PCTFE has a hexagonal unit cell having dimensions of a_o = 6.5 Å and C = 35 Å (15). A long fiber repeat distance of 43 Å indicates that the chain molecules are helices with a 14 - 16 monomer repeat cycle in every 360° of twist (16, 17). The fact that internal friction (logarithmic decrement) in PCTFE varies with density and temperature was established by Mc Crum (18). A detailed analysis of α, β, and γ relaxations by Hoffman et al. determined the extent of the effect of crystallinity on these dynamic transitions (19).

In contrast, PTFE is considered a highly crystalline (93 - 98%) polymer (20). The crystalline melting point is 327°C as compared to 218°C for PCTFE. It has a crystalline density of 2.30 g/cm^3. X-ray data show that the polymer exists in two helical conformations which at about 20°C consists of a 15 monomer repeat distance in each 180° of twist (21).

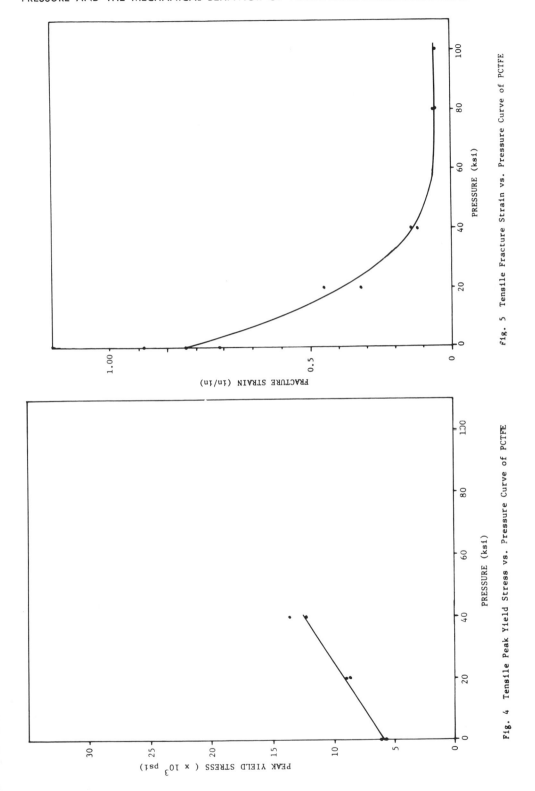

Fig. 5   Tensile Fracture Strain vs. Pressure Curve of PCTFE

Fig. 4   Tensile Peak Yield Stress vs. Pressure Curve of PCTFE

Several pressure-induced crystalline phase transitions, at
ambient temperatures, have been mentioned to occur in PTFE.
An internal friction study by Mc Crum showed that four
relaxation regions at 176°K, 300°K, 400°K, and 580°K occur
in PTFE which are dependent on temperature and crystallinity
(22). At atmospheric pressure, PTFE has a tensile elastic
modulus of 0.58×10^5 psi., a 2% offset tensile yield stress
between 2000 and 5000 psi., and a tensile elongation between
200 and 500%.

The experimental data show that four general pressure-
induced effects on the mechanical behavior of PCTFE are
clear: (1) a linear increase in compressive modulus from
1.69×10^5 psi. at atmospheric pressure to 4.37×10^5 psi. at
100,000 psi., with a discontinuity near 60,000 psi., (2) a
linear increase in the tensile peak yield stress from 5460
psi. at atmospheric pressure to 12,885 psi. at 40,000 psi.,
and (3) a decrease in tensile fracture strain from 0.94 in/in
at atmospheric pressure to 0.05 in/in at 100,000 psi.

Similarly, for PTFE there is: (1) a marked increase in
the compressive modulus from 0.59×10^5 psi. at atmospheric
pressure to 3.6×10^5 psi. at 100,000 psi. with a discontinuity
near 80,000 psi. due to a phase transition, (2) an apparent
non-linear increase in tensile yield stress (2% offset) from
about 1300 psi. at atmospheric pressure to 7500 psi. at
100,000 psi., and (3) a decrease in the tensile fracture strain
from about 1.3 in/in at atmospheric pressure to 0.25 in/in at
100,000 psi. (3). The phenomenon of necking was not observed
in tensile tests but cold-drawing at atmospheric pressure was
present. In Fig. 6 the marked differences in the shape of
the nominal tensile stress-strain curves for PCTFE and PTFE at
various pressures may be compared.

To describe the increase in elastic modulus with pressure,
the approach originated by Murnaghan and applied by Birch will
be employed (23, 24). The Birch equation for the pressure
dependent modulus is

$$E = Eo \left[1 + \frac{P}{Eo} 2 (5 - 4\nu)(1 - \nu) \right] \qquad (1)$$

where E represents the pressure dependent elastic modulus, Eo
is the elastic modulus at atmospheric pressure, P is the
applied hydrostatic pressure, and ν is Poisson's ratio at
atmospheric pressure. This equation may be expressed in reduced
form as

$$E = Eo + MP \qquad (2)$$

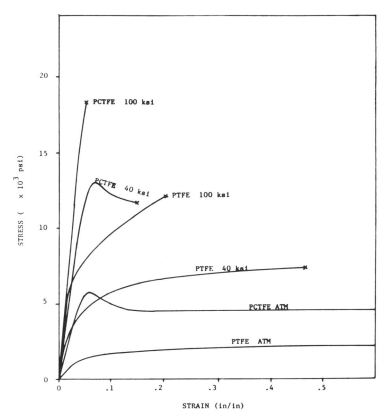

Fig. 6   Comparison of Nominal Tensile Stress-Strain Curves of PCTFE
and PTFE at Three Pressures

where the slope of the elastic modulus-pressure curve is
represented by

$$M = 2 (5 - 4\nu) (1 - \nu) \tag{3}$$

For PCTFE, Poisson's ratio is computed as 0.44, from the data
of Warfield et al. (25) which gives a predicted value for the
slope M of 3.6.   Thus, Eq. (2) becomes

$$E = E_o + 3.6 P \tag{4}$$

for PCTFE as compared to

$$E = E_o + 3.5 P \tag{5}$$

for PTFE.

Experimentally, the elastic modulus-pressure curve for
PCTFE has a lower linear region and an upper linear region.   The

lower region has a measured slope of 3.03 and the upper region
a slope of 2.08. In comparison, PTFE has an observed slope
of 3.30.

A molecular approach seems to offer a better qualitative
explanation for both the discontinuity in the modulus-pressure
curve and for the increase in modulus with pressure observed
for PCTFE. Consider the experiments mentioned above, concerning
internal friction, where certain relaxations were found to
depend on both temperature and crystallinity (18, 19). Herein
lies a possible clue to the origin of the discontinuity; if
one considers the fact that pressure-induced shifts of the glass
transition region have occurred in several polymers, then it
is reasonable to assume that relaxations which depend on
amorphous components can also be shifted to higher temperatures
(26).

It is normal for the dynamic modulus-temperature curves
to show a sharply decreasing slope while passing through
relaxation regions. Since pressure changes are analogous to
temperature changes, it is expected that a sudden change in
the slope of the elastic modulus-pressure curve can be viewed
as the shift of a low temperature relaxation process. In this
case, it is likely that the low temperature γ relaxation
located at about $-40°C$ is the one shifted upwards to room
temperature. The increase in elastic modulus of PTFE with
pressure may be partly due to the shifting of a low
temperature γ relaxation at about $-80°C$ up to room temperature
(3, 22).

The simplest approach to describing the linear increase
in tensile peak yield stress is through a modification of the
Mohr or Coulomb criterion (3). In modified form it becomes

$$\sigma_Y = \sigma_{T°} + kP \tag{6}$$

where σ_Y represents the peak yield stress at any pressure P,
$\sigma_{T°}$ is the value of axial yield stress in simple tension and k
is an experimentally determined constant obtained from the
slope of the tensile peak yield pressure curve. For PCTFE, k
has an observed value of 0.18 as compared to 0.08 for PTFE.
Thus the expression that describes simply the linear increase
of tensile peak yield stress with pressure for PCTFE data is

$$\sigma_Y = 5460 + 0.18 \, P \tag{7}$$

For PTFE, a comparable expression
$$\sigma_Y = 1300 + 0.08 \, P \tag{8}$$

is based on the 2% offset yield stress.

Ductility as observed by changes in the nominal tensile strain to fracture (Fig. 5), show that PCTFE proceeds from a rather ductile material to one which becomes extremely brittle with increasing pressure. At comparable pressure levels the ductility of PTFE exceeds that of PCTFE, namely, in elongation to fracture (compare data in Table 2 with that in Table 3). The PTFE samples did not exhibit necking but instead the cross sectional area of the gage length was reduced uniformly until fracture.

What kind of model will best explain what is happening within PCTFE while subjected to a deforming stress and hydro-static pressure? On the molecular level, the approach suggested by Hoffman et al. to describe the fine structure of bulk samples of PCTFE, will be adopted. Assuming that sample material was quick-quenched crystallized during the extrusion process, the polymer structure should consist of many very small crystallites. Each crystallite is composed of chain-folded lamellae which contribute toughness (low modulus) and ductility (high elongation) to the polymer. Anderson showed that molecules exceeding 1000 Å crystallize into chain-folded lamellae while molecules having lengths less than 1000 Å crystallize as extended chains (27, 28). It is noted that extended-chain crystals exhibit brittle behavior, namely, high elastic modulus and low elongation. PCTFE has molecules with number-average chain length of 9000 Å, so that chain-folded lamellae are more probable (19).

Crystallites are joined together by interlamellar tie molecules which enter and exit crystallites in a random manner. This is the switchboard reentry approach suggested by Flory and Mandelkern and called interlamellar amorphous model (29, 30). Basically, this model represents bulk-crystalline polymers as consisting of separate stacks of lamellae connected by amorphous areas. The presence of numerous long tie molecules in the amorphous regions are necessary to prevent extreme brittleness and to explain the rather good elastic properties found in PCTFE. Ductility, under a deforming stress, must depend primarily on the slipping and unfolding of chain molecules within the lamellae. High hydrostatic pressure restricts the motion of these tie molecules and unfolding of chain molecules within the lamellae. As a result, the amorphous regions become stiff and lose the ability to deform and absorb energy. Consequently, increasing brittleness with increasing pressure should be observed in semi-crystalline polymers like PCTFE.

TABLE 3

Average Values of Elastic Modulus and Tensile Fracture
Strain at Various Pressures for PTFE

Pressure (psi.)	Compressive Modulus (x 10^5 psi.)	Fracture Strain (in/in)
Atmospheric	0.59	1.27
20,000	1.16	.78
40,000	1.79	.47
60,000	2.65	.27
80,000	2.49	.32
100,000	3.60	.23

Admittedly, the interlamellar amorphous model gives no
attention to possible spherulitic structure in bulk crystallized
polymers and by doing so ignores the inclusion of inter-
spherulitic boundaries in the explanation of fracture behavior
(31). However, since the PCTFE used in this investigation
was quench-crystallized, the use of the interlamellar amorphous
model is plausible.

ACKNOWLEDGEMENTS

It is a pleasure to acknowledge the receipt of technical
assistance from Dr. Lynn Salisbury and Mr. S. K. Bhateja.

REFERENCES

1. C. E. Weir, J. Res. N.B.S., 53, 245, (1954).
2. W. R. Warfield, J. Appl. Chem. 17, 263, (1967).
3. J. A. Sauer, D. R. Mears and K. D. Pae, European Polymer
    J. 6, 1015 (1970).
4. K. D. Pae, D. R. Mears and J. A. Sauer, J. Polymer
    Sci., B, 6, 773 (1968).
5. D. R. Mears, K. D. Pae and J. A. Sauer, J. Appl. Phys.,
    40, 4229 (1969).
6. K. D. Pae and J. A. Sauer, Mech. Eng., The Institute of
    Mechanical Engineers, London (in Press).
7. D. Sardar, S. V. Radcliffe, and E. Baer, Polymer Eng'g.
    Sci., 8, 290 (1968).

8. W. I. Vroom and R. F. Westover, SPE J., 25, 58 (1969).
9. H. Ll, D. Pugh, E. F. Chandler, L. Holliday and J. Mann, Polymer Eng'g. Sci., 11, 463 (1971).
10. D. R. Mears, Ph.D. dissertation, Rutgers University, 1968.
11. Kel-F 81 Brand Plastic, Technical Information, Chemical Division, 3M Company.
12. J. D. Hoffman and J. J. Weeks, J. Research Nat. Bur. Standards, 60, 465 (1958).
13. R. P. Kambour, Polymer Eng'g. Sci., 8, 281 (1968).
14. K. D. Pae and D. R. Mears, J. Polymer Sci., B, 6, 269 (1968).
15. H. S. Haufman, J. Am. Chem. Soc., 75, 1477 (1953).
16. C. Y. Liang and S. Krimm, J. Chem. Phys., 25, 563 (1956).
17. G. V. D. Tiers and F. A. Bovey, J. Polymer Sci., A, 1, 833 (1963).
18. N. G. Mc Crum, J. Polymer Sci., 60, S 3, (1962).
19. J. D. Hoffman, G. Williams and E. Passaglia, J. Polymer Sci., C 14, 173 (1966).
20. F. W. Billmeyer, Textbook of Polymer Science. Wiley, New York (1971).
21. P. H. Geil, Polymer Single Crystals Interscience, New York (1963).
22. N. G. Mc Crum, J. Polymer Sci., 34, 355 (1959).
23. F. D. Murnaghan, Finite Deformations of an Elastic Solid. Wiley, New York (1951).
24. F. Birch, J. Appl. Phys. 9, 4 (1938).
25. R. W. Warfield, J. E. Cuevas, and F. R. Barnet, J. Appl. Poly. Sci., 12, 11 47 (1968).
26. J. M. O'Reilly, J. Polymer Sci., 57, 429 (1962).
27. F. R. Anderson, J. Appl. Phys. 35, 65 (1964).
28. F. R. Anderson, J. Polymer Sci. C 3, 275 (1963).
29. P. J. Flory, J. Amer. Chem. Soc. 84, 2857 (1962).
30. L. Mandelkern, J. Polymer Sci., C 15, 129 (1966).
31. G. C. Oppenlander, Science, 159, 3821 (1968).

TRANSPORT AND MECHANICAL PROPERTIES OF POLYCARBONATE

J.A. Eilenberg and W.R. Vieth

Department of Chemical and Biochemical Engineering

Rutgers University, New Brunswick, New Jersey

ABSTRACT

Sorption kinetics and equilibria for CO_2, A, and CO in poly-carbonate were studied over a range of temperatures from 25°C to 120°C (Tg $\simeq$ 150°C) and pressures from 3 to 23 atm. abs. The dual mode sorption model of Vieth and Sladek (1) was used to test the data. The microvoid free volume fraction was found to be the same as that for glassy polystyrene (2).

Stress relaxation studies were conducted in the linear visco-elastic regime (strain levels < 1.5%) below Tg. Reduction of data to 25°C, 85°C, and 105°C and analysis of the master curves by Tobolsky's procedure (3) produced calculated primary relaxation times significantly lower than expected for a glassy polymer.

The observed activation energy for diffusion is approximately one-half that expected for an activated transport mechanism. The level of fractional microvoid (free) volume, coupled with the sec-ondary relaxations known for polycarbonate, implicate chain seg-mental micromotions and slip flow in low density, amorphous regions as important factors in the transport of gas in the polymer system.

INTRODUCTION

The Engineering Plastics may well be considered to be those polymers which are replacing, on a large scale, such natural ma-terials as wood, sand (glass), and metals. The polymethacrylates, polystyrenes, and polycarbonates with glass transition temperatures

145

greater than 25°C have found extensive use in toys, lenses for
automotive and recreation vehicles, and in packaging and glazing.
Of prime importance to the selection of a material for a particular end use are its transport and mechanical properties. The use
of suitable small gas molecules as probes (4) and mechanical testing to determine moduli and relaxation times can provide a coher-
ent picture of them.

    Bisphenol A polycarbonate (Lexan)* was the material of choice
for this investigation. An unusually high reported impact strength
(12-17.5 ft-lb/in. Izod) was an unexplained phenomenon providing a
primary motivation for further study of this particular glassy
polymer.

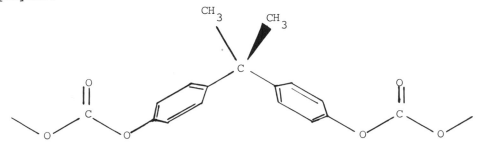

Fig. 1.   Polycarbonate backbone

In Figure 1 the polycarbonate backbone is seen to contain two
stiff aromatic rings. However, some flexibility is introduced by
the presence of oxygens even though the carbonate group, as a
whole, is highly polar. The methyl groups also contribute some
low-amplitude motion to the backbone 70°C below the glass transition temperature (5). It was postulated that this behavior may
be partly responsible for the high observed impact strength.

    The fundamental tool for this research was the Dual-Mode Sorp-
tion Model (DMSM)** developed and tested by Vieth, et. al. (6,4,7).

$$C = k_D P_f + \frac{C'_H b P_f}{1+b P_f} \tag{1}$$

---

* General Electric registered trademark

** Author's notation

C = solubility, c.c.(STP)/c.c. polymer

k_D = Henry's Law dissolution constant,
c.c.(STP)/c.c. atm.

P_f = pressure, atm.

C_H' = microvoid saturation constant, c.c.(STP)/c.c.
polymer

b = microvoid affinity constant, atm^{-1}

The model allows analysis of observed nonlinear isotherms for individual sorption coefficients and, especially, permits estimation of microvoid free volume fraction. The first mode of sorption is considered to be a linear Henry's Law relationship while the second mode is described by a Langmuir nonlinear term. This second mode describes the filling of microvoids which immobilize a portion of the gas molecules. These microvoids are frozen into the polymer matrix as the material traverses the glass transition temperature and loses its rotational degrees of freedom. As regards mechanical properties, the level of the volume fraction of microvoids (free volume) can be of greater significance than their individual size and distribution.

Stress relaxation data gathered in this work and by Yannas, et. al. (8) were examined for both moduli and relaxation times as a function of temperature and strain level in the linear viscoelastic region. In this region the polymer exhibited Hookian stress-strain and Newtonian viscous behavior. For polycarbonate this was found to be applicable to strain levels near 1% (8). The relaxation times for materials below their glass transition are decidedly longer (> 10^6 sec) than was practically feasible to run. Time-temperature superposition, TTS*, allows data taken at different temperatures; but at constant strain or stress, to be reduced to one temperature covering orders of magnitude of time. Williams, Randel, and Ferry (9) examined a variety of amorphous, glass-forming substances with this technique in 1955. They wished to examine the effect of temperature on the characteristic relaxation time of a material. However, by application of their WLF equation to stress relaxation data fractional free volumes for materials may be determined.

Another approach to the reduction of stress relaxation data was proposed by Tobolsky and Murakami (3) in 1959. Instead of assuming a single relaxation time as in the WLF relation a discrete distribution was proposed.

_____

* Author's notation

$$E_r(t) = E_a \exp\left(-\frac{t}{\tau_a}\right) + \cdots + E_{m-1} \exp\left(-t/\tau_{m-1}\right)$$

$$+ E_m \exp\left(-t/\tau_m\right) \tag{2}$$

$$E_r(t) = \text{relaxation modulus, dynes/cm}^2$$
$$\tau_m = \text{primary relaxation time, sec.}$$
$$m = \text{primary}$$
$$m-1 = \text{secondary}$$

Using this approach, the primary relaxation times for polycarbonate determined by this technique were quite interesting as will be discussed further in a later section.

APPARATUS, PROCEDURES, AND MATERIALS

The high pressure sorption apparatus shown in Fig. 2 was originally built by Edward Ma (13) at MIT in 1964. Vieth, et. al. (7) and other investigators (2, 11) used and improved this equipment to verify the DMSM. Though the system was capable of sorption studies using an initial pressure of 500 psig, the pressure levels for this examination were carried to ca. 350 psig. The characteristic isotherms were well illustrated at these lower pressures (Fig. 3).

The sample bomb had a nominal capacity of 300 c.c. and 49.90 c.c. of polycarbonate strips were placed within it. The bath for this bomb maintained temperature control to ± 0.1C°. The system was evacuated to 0.02 mm Hg and the system was then isolated from the vacuum pump by closing valve PV7. The gas cylinder valve was opened to build up pressure at PV1. Valve PV1b was cracked open, pressurizing the bomb, until the desired pressure level was reached, as seen on the Crosby pressure gauge. Having then closed this valve, the following measurements were noted: initial pressure, initial ambient temperature, and bath temperature. Upon reaching equilibrium, the final pressure and temperature were recorded, the system was evacuated through PV6, and the vacuum pump was used to bring the system to 0.02 mm Hg once again. A typical isotherm was generated in 50 psig increments.

The mass balance for the system is,

gas in = gas sorbed + gas in void space

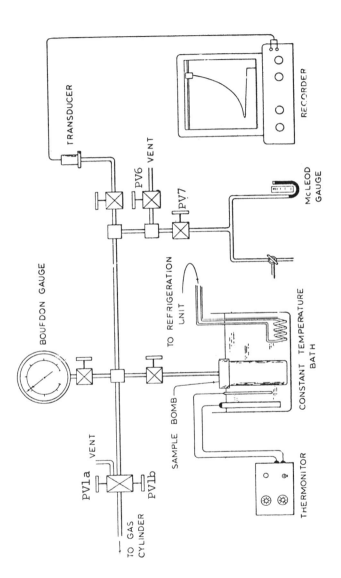

Fig. 1. Schematic diagram of the high pressure sorption apparatus

$$\frac{V_{ai}P_i}{z_{ai}T_{ai}} + \frac{V_b P_i}{z_{bi}T_b} = \frac{CV_p P_s}{z_s T_s} + \frac{V_a P_f}{z_{af}T_{af}} + \frac{V_b P_f}{z_{bf}T_b} \tag{3}$$

$$C = \frac{z_s T_s}{V_p P_s}\left[V_a\left(\frac{P_i}{z_{ai}T_{ai}} - \frac{P_f}{z_{af}T_{af}}\right) + \frac{V_c}{T_b}\left(\frac{P_i}{z_{bi}} - \frac{P_f}{z_{bf}}\right)\right] \tag{4}$$

C = solubility, c.c.(STP)/c.c. polymer
z = compressibility
P = pressure, atm
T = temperature, °K
V_p = volume of polymer sample, 49.90 c.c.
V_a = void volume of system at ambient conditions,
        36.0 c.c.
V_b = void volume of system at bath temperature,
        249.88 c.c.

subscripts

i = initial
f = final
a = ambient
b = bath
s = standard

Substituting the values given into the equation yields

$$C = \frac{1366.914\, z_s}{T_b}\left(\frac{P_i}{z_{bi}} - \frac{P_f}{z_{bf}}\right) + 196.921\left(\frac{P_i}{z_{ai}T_{ai}} - \frac{P_f}{z_{af}T_{af}}\right) \tag{5}$$

To determine diffusion constants the final run histories for each isotherm were also followed; i.e., the pressure decay was noted as a function of time. The linear isotherms of A and CO allowed convenient application of the Crank and Park "half-time" solution (12).

$$D = 0.0492\left(\frac{L^2}{t_{1/2}}\right) \tag{6}$$

with the conditions

$$t = 0 \qquad\qquad C_D = 0 \qquad\qquad -\ell \le x \le \ell$$

$$t \ge 0 \qquad\qquad \frac{\partial C_D}{\partial x} = 0 \qquad\qquad x = 0$$

$$t > 0 \qquad\qquad C_D = k_D P \qquad\qquad x = \pm\ \ell$$

$$\ell = \text{half-thickness of film}$$

The diffusion constant was obtained using a scaling factor to superpose data onto a normalized solution generated by Vieth and Sladek (13).

The mechanical testing was done on the Instron tensile tester with an environmental chamber supplied by the same company. The Instron is a highly accurate device capable of position within ± 0.001" and the environmental chamber held temperatures to ± 1.0F°. The samples (1.750" x 0.264" x 0.030") were subjected to strains from 0.29% to 1.43%; which, depending on temperature, lie in the linear viscoelastic regime for polycarbonate. These stress relaxation experiments ran for one hour each.

RESULTS AND DISCUSSION

The isotherms for CO_2, A, and CO are shown in Figs. 3, 5, and 6. Henry's Law was obeyed by A and CO over the entire range of pressures studied. In Fig. 7 the natural logarithm of the solubility constant, k_D, (Table 2A) is plotted as a function of reciprocal temperature. The resulting enthalpies of sorption, ΔH_s, (Table 2B) show that CO_2 had a stronger tendency to interact with polycarbonate than A or CO. The gas force constant, ε/k, may be used to characterize the dissolved gas molecule-polymer interaction. It would be expected that ΔH_s would become more negative with increasing ε/k. This was not entirely true for the gases studied.

TABLE 1A   SOLUBILITY CONSTANTS FOR CO_2, A, AND CO IN POLYCARBONATE

T	$T^{-1}(\times 10^3)$	$k_D(CO_2)$	$k_D(A)$	$k_D(CO)$
25	3.36	1.28	0.36	0.31
45	3.14	1.01	0.29	0.20
65	2.96	0.58	0.21	0.15
85	2.79	0.39	0.15	0.12
105	2.65	-	0.13	0.08
120	2.55	-	0.11	0.07

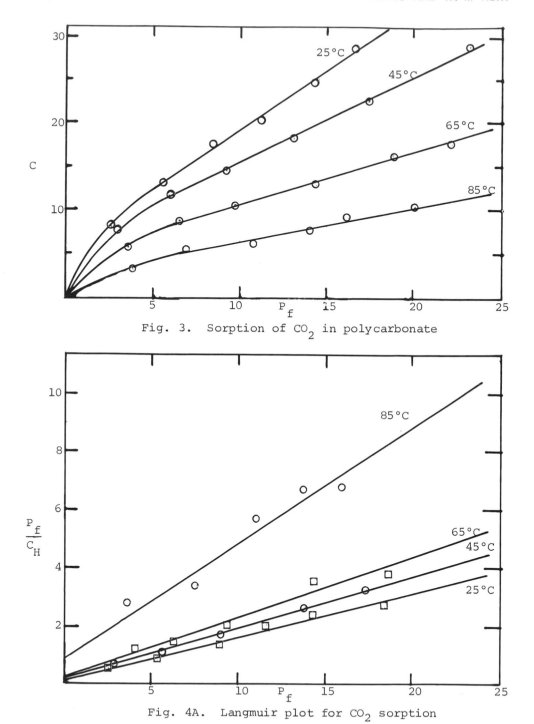

Fig. 3.   Sorption of CO_2 in polycarbonate

Fig. 4A.   Langmuir plot for CO_2 sorption

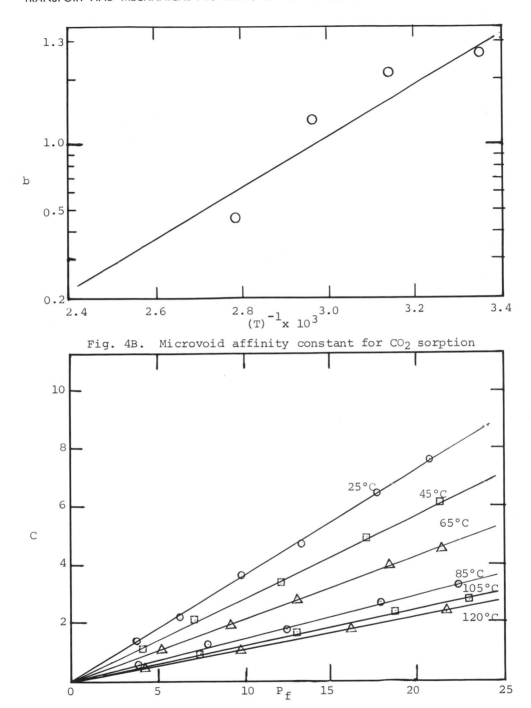

Fig. 4B. Microvoid affinity constant for CO_2 sorption

Fig. 5. Argon sorption in polycarbonate

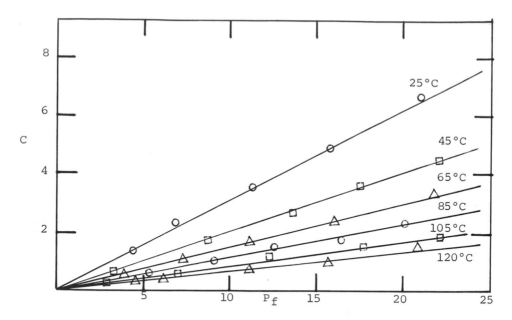

Fig. 6.   Carbon monoxide sorption in polycarbonate

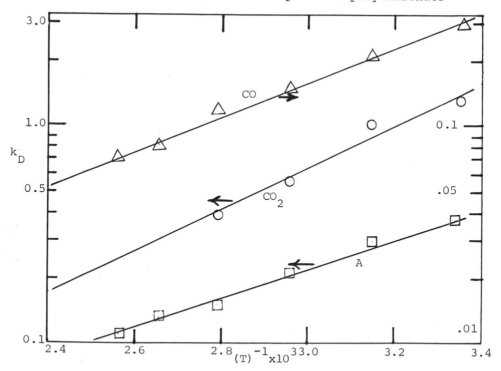

Fig. 7.   Solubility constants of gases in polycarbonate

TABLE 1B   ENTHALPIES OF SORPTION

Gas	ΔH_s (kcpm)	ε/k
CO_2	-4.2	229
A	-3.0	122
CO	-3.7	110

The dual-mode parameters for CO_2 shown in Table 2A are derived from the slope and intercept of the Langmuir plot (Fig. 4A).
As is shown in the isotherms, CO_2 is much more soluble in polycarbonate than the other gases. It may well be that the highly polar
carbonate group induces dipole interactions with the CO_2 molecule.
This, together with the high ε/k and observed heat of sorption,
could well account for these results.

TABLE 2A   DUAL-MODE SORPTION PARAMETERS
FOR CO_2 IN POLYCARBONATE

T	$T^{-1}(\times 10^3)$	k_D	C_H'	θ_{mv} (%)	b
25	3.36	1.28	6.48	1.40	2.26
45	3.14	1.01	5.55	1.20	2.16
65	2.96	0.58	4.94	1.07	1.30
85	2.79	0.39	2.57	0.56	0.47

TABLE 2D   DUAL-MODE SORPTION PARAMETERS FOR
CO_2 IN UNORIENTED POLYSTYRENE (2) *

25	3.36	0.90	6.52	1.41	0.26

The diffusion constants for CO_2 are determined by the curve-
matching procedure alluded to earlier. Values determined by this
technique and the half-time solutions for A and CO are shown in
Table 3A. Relative to other polymers, these gas transport coefficients are rather large. Though the sorption level is quite high
the diffusivities for CO_2 are lower by nearly an order of magnitude than those for A and CO. The activation energies for diffusion, E_D, shown in Table 3B, indicate a modified process; i.e., the
values are significantly less than the expected 10-12 kcpm for a
totally activated process. Slip-flow contributions to the diffu-
sional process in low density regions of the polymer are strongly
indicated.

---

* Reported values corrected by authors for compressibility considerations.

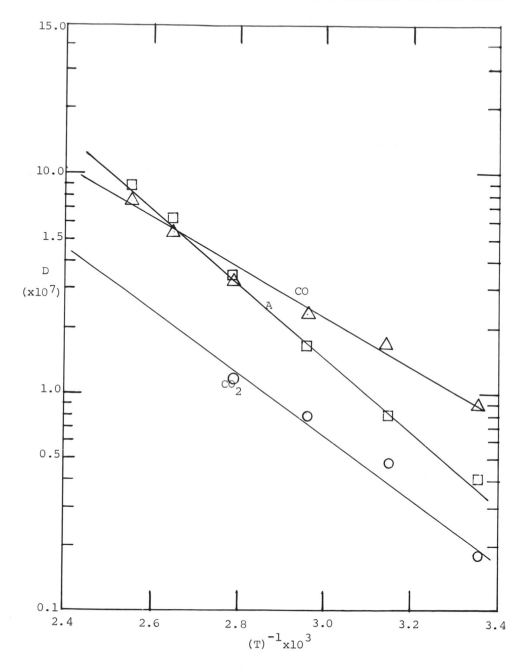

Fig. 8.   Diffusion constants for gases in polycarbonate

TABLE 3A   DIFFUSION CONSTANTS FOR CO_2, A, CO IN POLYCARBONATE

T	$T^{-1}(x10^3)$	$D_{CO_2}(x10^8)$	$D_A(x10^7)$	$D_{CO}(x10^7)$
25	3.36	1.73	.40	.86
45	3.14	4.70	.76	1.67
65	2.96	7.71	1.59	2.23
85	2.79	11.52	3.40	3.25
105	2.65	-	6.16	5.43
120	2.55	-	8.93	7.52

TABLE 3B   ACTIVATION ENERGY FOR DIFFUSION

Gas	$d^2(x10^{16})$	$E_D(kcpm)$	$E_D(kcpm)$ (22)
CO_2	16.4	6.58	9.0
A	13.0	7.84	6.0
CO	13.0	5.08	-

d^2 = (equivalent molecular diameter)^2

The sorption process is seen as microvoid-filling accompanied by normal dissolution. The molecular periphery of the microvoids could be made up of polar carbonate groups. Their interaction with CO_2 molecules could have the effect of detaining additional molecules rather tenaciously. The apparent enthalphy of hole-filling (ref. Fig. 4B) is 5.31 kcpm while for CH_4 in oriented PS is only 3.60 kcpm (7). This is evidence of the tenacity of this second mode. As the gases in this study were all of the same general size, this would seem a logical explanation for the order of magnitude difference in the diffusion constant. The slightly larger size of CO_2 also contributes, no doubt.

Of great interest is the magnitude of C_H', which is a direct measure of microvoid free volume.

$$\theta_{mv} = C_H' * (mol.\ wt.\ of\ gas) * (molar\ vol.\ at\ STP)^{-1} * (liq.\ density\ at\ STP)^{-1}$$

$$\theta_{mv} \simeq (2.17 \times 10^{-3})\ C_H'\ for\ CO_2 \qquad (8)$$

Because of the assumption that the sorbed molecules are packed into a liquid-like lattice, the corresponding estimates of the free volume fraction are the minimum values. The actual sorbate density is probably somewhat lower and the true free volume fractions correspondingly higher. As is seen in Tables 1A & B the resultant values of microvoid (free) volume percentage for CO_2 in polycarbonate were of the same magnitude as for CO_2 in unoriented polystyrene (20), another glassy material with high impact resistance. In the glassy state a small increase in microvoid free volume fraction may significantly alter the ability of chain segments to move. The transition from rubbery to glassy freezes the microvoids into the matrix by the disappearance of the rotational modes of freedom (14). However, the glassy state is a nonequilibrium one; i.e., segments are continually undergoing slow relaxation. Results (15-20) from previous dynamic testing of polycarbonate show that this material has several secondary relaxations below Tg. Though the explanations for what is "relaxing" are conflicting, the fact remains that the constituent molecules of the polycarbonate chain are undergoing reorientation movement at room temperature and above.

The results for mechanical testing, normalized to 25°C, 85°C, and 105°C are shown in Figs. 9-11. The experiment was extremely temperature sensitive as is seen in the plot reproduced in Fig. 12. The environmental chamber variations of ± 1.0 F° were too great for obtaining absolutely smooth relaxation curves. However, normalizing the stress relaxation results produces smooth master curves. When Tobolsky's procedure (3) is applied to the normalized stress relaxation results, it is found to yield primary relaxation times near the rubbery domain. Table 4 presents the results of this procedure, as well as an analysis by the authors of data obtained by Yannas, et. al. (8) at 85°C and 100°C.

TABLE 4   PROCEDURE X PRIMARY RELAXATION
TIMES, τ_m (sec)

T	$\tau_m (\times 10^{-6})$	$\tau_m (\times 10^{-6})$ (8)	$\tau_m (\times 10^{-6})$ (3)
25	5.48	–	–
85	1.54	1.40	–
105	0.63	0.32	–
115	–	–	0.27*

* For polydisperse Polystyrene (B) with $\overline{M}_w/\overline{M}_n$ = 1.5, which is

narrow for a polydisperse polymer, in the rubbery flow regime (115°C).

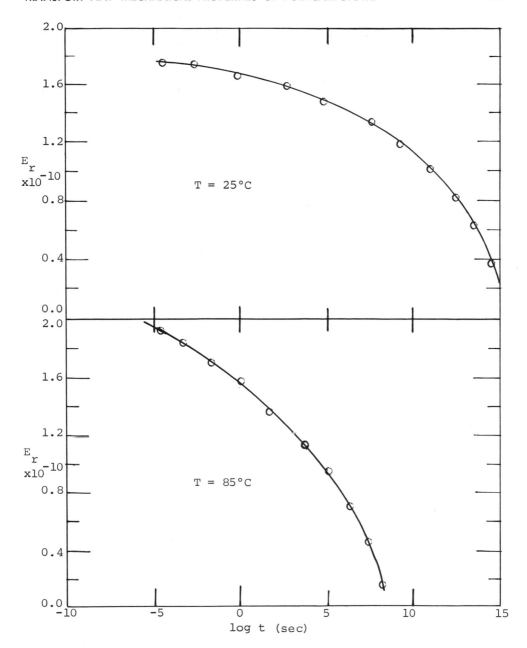

Figs. 9-10.   Stress relaxation master curves for polycarbonate

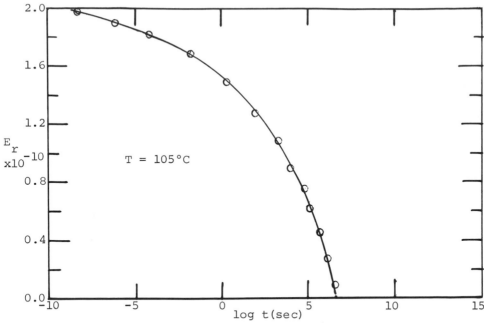

Fig. 11.   Stress relaxation master
curve for polycarbonate

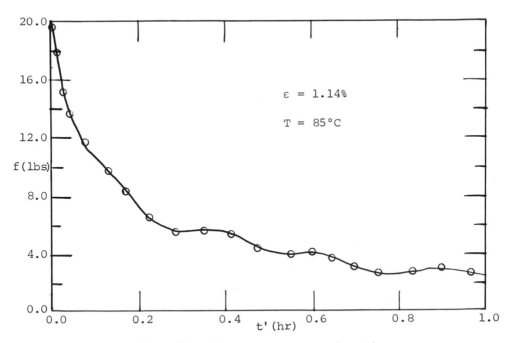

Fig. 12.   Instron stress relaxation
raw data chart output

The relaxation times for 25°C may be considered to be at the very low regime of glassy behavior while the results at higher temperatures are most definitely in the rubbery regime. The highest temperature examined by this procedure is still 44 C° below Tg. The energy is being dissipated in the microvoids and diffusive pathways for chain segments (or gas molecules) by these relaxations. This result is very significant and reinforces the explanation for the observed high impact strength of polycarbonate.

CONCLUSION

The DMSM provides a quite general method for describing sorption kinetics and equilibria for gases in glassy polymers. Of particular interest is the evaluation of microvoid free volume fraction. Gas molecular probes; particularly CO_2, show there is adequate space for movement below Tg. Dynamic tests show there are several secondary relaxations occuring below Tg. Indeed, examination of stress relaxation data indicate there is freer movement of chain segments in polycarbonate than is usual for glassy polymers.

ACKNOWLEDGMENT

The authors wish to express their gratitude to the National Science Foundation (NSF-GK 14075), whose financial support was instrumental in the accomplishment of this work.

BIBLIOGRAPHY

1.   Vieth, W.R., and Sladek, K.J., J. Colloid Sci., 20, 9 (1965).

2.   Tam, P.M., "High Pressure Sorption in a Glassy Polymer," S.M. Thesis, M.I.T., Cambridge, Massachusetts (1965).

3.   Tobolsky, A.V., and Murakami, K., J. Polymer Sci., Vol. XL, 443 (1959).

4.   Vieth, W.R., "A Study of Poly(ethylene terephthalate) by Gas Permeation," Sc.D. Thesis, M.I.T., Cambridge, Massachusetts (1961).

5.   Yannas, I.V., J. Macromol. Sci. - Phys., B6(1), 91 (1972).

6.   Michaels, A.S., Vieth, W.R., and Barrie, J.A., J. Appl. Phys., 34, 1 (1963).

7.   Vieth, W.R., Frangoulis, C.S., and Rionda, J.A., Jr., J.
     Colloid and Interface Sci., 22, 454 (1966).

8.   Yannas, I.V., Sung, N-H., and Lunn, A.C., J. Macromol. Sci.-
     Phys., B5(3), 487 (1971).

9.   Williams, M.L., Landel, R.F., and Ferry, J.D., J. Am. Chem.
     Soc., 77, 3701 (1955).

10. Ma, E., "Construction and Operation of a High Pressure Sorp-
     tion System," Chem. Eng. Dept., M.I.T. (1964).

11. Wuerth, W.F., "A Study of Polypropylene by Gas Permeation,"
     Sc.D. Thesis, M.I.T. (1967).

12. Crank, J., and Park, G.S., eds., "Diffusion in Polymers,"
     Academic Press, London, p. 16 (1968).

13. Loc. cit., p. 1027.

14. Meares, P., J. Am. Chem. Soc., 76, 3415 (1954).

15. Reding, F.P., Faucher, J.A., and Whitman, R.D., J. Poly. Sci.,
     54 556 (1961).

16. Heijboar, J., J. Poly. Sci. Pt C, No. 16, p. 3755 (1968).

17. Locati, G., and Tobolsky, A.V., Advan. Md. Relaxation Pro-
     cesses, 1, p. 375 (1970).

18. Garfield, L.J., J. Poly. Sci., Pt C, No. 30, p. 551 (1970).

19. Yannas, I.V., Polymer Letters, 9, p. 611 (1971).

20. Chung, C.I., and Sauer, J.A., J. Poly. Sci. Pt A-2, 9,
     p. 1097 (1971).

21. Loc. cit. p. 450.

22. Norton, F.J., J. Appl. Poly. Sci., 7, p. 1650 (1963).

VISCOELASTIC AND DIELECTRIC PROPERTIES OF A-B-A TYPE

BLOCK COPOLYMERS

R. T. Jamieson, V. A. Kaniskin*, A. C. Ouano** and M. Shen
Department of Chemical Engineering
University of California
Berkeley, California 94720

ABSTRACT

The viscoelastic and dielectric properties of two A-B-A type block copolymers were determined. One contains styrene-butadiene-styrene (SBS) blocks and the other styrene-isoprene-styrene (SIS) blocks. Samples were prepared by casting from solutions of these polymers in benzene/heptane, carbon tetrachloride and methylethyl-ketone/tetrahydrofuran. The viscoelastic behavior of SIS block copolymers is apparently unaffected by the casting solvent, while that of the SBS block copolymers is similar from the former two solvents but drastically modified for the MEK/THF-cast sample. Shift factors utilized in constructing the viscoelastic master curves were found to obey the WLF equation as modified by Shen and Kaelble, except for the MEK/THF-cast SBS block copolymer. Dielectric loss curves were obtained over an extended frequency range. For comparison, the mechanical loss curve for a CCl_4-cast SBS block copolymer was computed from master curve by linear programming. In both cases the only observed loss peaks were those resulting from the primary glass transitions of the respective blocks. No peak was found in the intermediate temperature range that may be attributable to the "mixed" regions.

*International Research Exchange Scholar (1970-1971), Permanent Address: Leningrad Polytechnical Institute, U.S.S.R.

**IBM Research Laboratories, San Jose, California.

INTRODUCTION

In recent years considerable interest has developed in a new
class of macromolecules -- triblock copolymers. The molecules have
an A-B-A structure where B is a rubbery block (such as polybutadiene
or polyisoprene) bounded by A blocks which are glassy in nature
(e.g., polystyrene). These new materials are important from an
engineering viewpoint because they behave as vulcanized rubbers at
room temperature but can be processed as thermoplastics at higher
temperatures. In addition, they have shown enhanced tensile
properties over those of ordinary elastomers.

The unique properties of A-B-A copolymers are a result of the
thermodynamic incompatibility of the component blocks which gives
rise to phase separation : for example, the polystyrene ends will
form into glassy domains anchoring the central polybutadiene or
polyisoprene chain. These domains then act as physical crosslinks
as well as filler, thus imparting excellent elastomeric properties
to the material.

Abundant experimental evidence has supported the multiple phase
morphology and a number of theoretical treatments based on statistical thermodynamics have appeared.[1-3] One of the most interesting
features of block copolymers is that the morphological structure of
a given sample may be changed by casting from different solvents.
It is thus of interest to investigate the effect of this supramol-
ecular structure on the physical properties of these polymers. In
this work the viscoelastic and dielectric behavior of two A-B-A
types of block copolymers cast from a number of solvents will be
reported.

1. Materials

Two polymers were used in this study -- Kraton 1101 and Kraton
1107, both supplied by the Shell Chemical Company. Kraton 1101 is
a styrene-butadiene-styrene block copolymer (SBS) with number
average molecular weight 76,000 and containing 28% by weight styrene.
The polybutadiene portion consists of 51% trans - 1,4, 41% cis -
1,4, and 8% vinyl structure.[4] Kraton 1107 is another triblock
copolymer but with polyisoprene as the central block (SIS). The
fraction of polystyrene in Kraton 1107 was determined by infrared
and nuclear magnetic resonance spectroscopy. Figure 1 shows the
ir spectra, taken by a Perkin-Elmer Model 21 Spectrometer, for the
pure polystyrene, pure polyisoprene and Kraton 1107 in 2.5% carbon
tetrachloride solutions. The 14.3μ absorption peak, which is characteristic of polystyrene,[5] was used to determine the composition
of Kraton 1107. It was found that the latter contains 10% polystyrene. The nmr spectrum for Kraton 1107 (5% CCl_4 solution) is

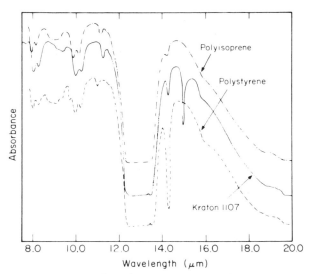

Fig. 1 Infrared spectra of polyisoprene, polystyrene, and Kraton
1107 (SIS block copolymer).

given in Figure 2. A Varian A60 NMR Spectrometer was used in the
analysis. By analyzing the intensity of the 7.1 ppm peak, characteristic of the phenyl group protons[6], the polystyrene content was
found to be 8%. Thus the ir and nmr results are in fair agreement
with each other.

The molecular weight and molecular weight distribution of Kraton
1107 were determined by gel permeation chromatography. The Water's
Associates GPC Model 200 with automatic sample injection and modified
to allow computer interfacing was used. The analog signal from the
differential refractometer was directly fed to an IBM 1800 computer
with an analog to digital converter and random access data storage
system. At the end of the analysis, the chromatographic data were
reduced to a normalized molecular weight distribution curve, and to
the different averages of the molecular weight distribution by the
computer with minimum personal intervention.[7] Figure 3 shows the
normalized chromatogram for Kraton 1107. The number average molecular weight of Kraton 1107 was found to be 1.74×10^{5}, and the hetero-
geneity index 1.34.

Samples were prepared as follows. A 10% solution of the polymer
was made with either carbon tetrachloride, benzene containing 10%
by volume heptane, or tetrahydrofuran containing 10 % by volume
methylethylketone. The solution was filtered, and then poured onto
the surface of mercury. After approximately two weeks, the sample
was dried in a vacuum oven at 40-45°C until no change in weight was
observed. Samples for viscoelastic testing were approximately 1.5 mm

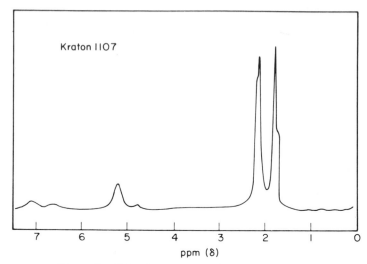

Fig. 2   NMR Spectrum of Kraton 1107.

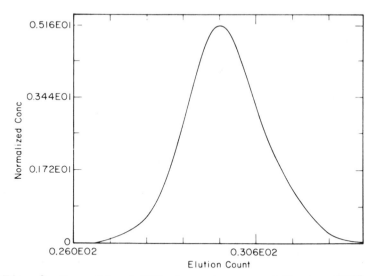

Fig. 3   Normalized GPC chromatogram of Kraton 1107.

thick, and were cut into 6.5 cm × 5 cm strips on a milling machine using a circular knife edge. Dielectric samples were cast 0.01 cm thick and 5 cm in diameter.

2. Viscoelastic Measurements

Stress relaxation measurements were made on a Model TM-SM Instron Universal Testing Instrument equipped with a custom-built environmental chamber. Temperature was controlled to ±0.5°C by a Missimers Model PITC Temperature Conditioner. Vertical distances were measured with a cathetometer accurate to ±.005 cm. Stress relaxation was carried out in the tensile mode until the modulus reached 10^9 dynes/cm;^{2} above this value the flexural mode was used.

In tensile testing, two flat rectangular pieces of aluminum were affixed to each end of the sample with epoxy cement. These end pieces contained holes so they would fit onto lugs on special devices inserted in the grips. Thus the sample could be held securely without being deformed at the ends.

The experimental procedure was as follows. A stress-strain curve was first obtained for each new set of samples to check for yielding at low strains. A crosshead speed of 2 cm/min was used. None of the polymer-solvent combinations shows yielding. For modulus-time measurements, the chamber was first brought to the desired temperature. The sample was stretched between 3 and 10% and allowed to relax up to 4000 sec. In flexural relaxation strains were kept below 3%. Sample width and thickness were corrected for temperature using a linear coefficient of expansion of .0002 $(°C)^{-1}$ for the nonglassy and .0001 $(°C)^{-1}$ for the glassy state. These are typical average values for many polymers.

3. Dielectric Measurements

Dielectric constant ε' and loss ε'' for the block copolymers were determined over the frequency range of 50 Hz to 10^5 Hz. A General Radio Model 1615A capacitance bridge and a 1232A null detector were used. Signal source was a Hewlett-Packard oscillator. The sample was inserted in the electrode, which was then placed in a Statham SD30 Temperature Test chamber. Temperature was maintained constant to ±0.5°C while capacitance and dissipation measurements were made over the entire frequency range. Temperature was then reset to cover the range between -160°C to +120°C for all samples.

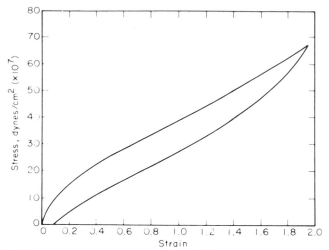

Fig. 4 Stress-strain curve for CCl_4 cast SBS at 25°C.

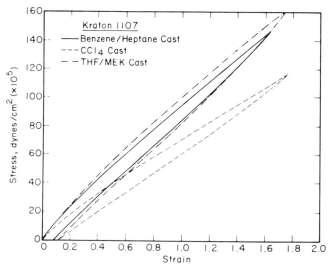

Fig. 5   Stress-strain curves at 25°C for SIS cast from various solvent systems.

RESULTS AND DISCUSSION

1. Stress-Strain Behavior

Figures 4 and 5 show the stress-strain curves for SBS and SIS cast from the various solvent systems. The most interesting features are that there is no yielding for the THF/MEK cast SIS and that its initial modulus is nearly equal to that for the benzene/heptane cast sample. These findings are significant when compared to the results of Beecher et al.[4] for SBS. They found that samples cast from THF/MEK show yielding at 3% strain and have a substantially higher initial modulus than samples cast from benzene/heptane. Their interpretation, supported by electron micrographs, is that MEK, a better solvent for polystyrene than for polybutadiene, evaporates last, swelling the polystyrene domains. These swollen domains come into contact, forming frequent interconnections among polystyrene domains. The interconnections are what give the THF/MEK cast sample its high initial modulus, and subsequent yielding at higher strains. Beecher et al. also studied several SIS copolymers cast from THF/MEK. They found that yielding depends upon the weight fraction of polystyrene and the concentration of MEK -- the higher either one of these quantities, the greater the likelihood for interconnections to form between styrene domains. The stress-strain results in Figure 5 thus seem to indicate that the THF/MEK cast SIS as prepared for this work is polyisoprene-continuous.

In casting SIS from benzene/heptane, the benzene will evaporate first to leave a heptane swollen polyisoprene phase, heptane being a good solvent for polyisoprene and a poor solvent for polystyrene. The swollen phase will isolate the polystyrene domains giving a polyisoprene-continuous morphology. Since THF/MEK and benzene/heptane cast SIS have very similar stress-strain curves, they probably have similar morphologies as well. Interconnections among polystyrene domains are probably minimal, due to the low styrene content in SIS. This fact is demonstrated in the electron photomicrograph of a comparable sample recently.[8] The SIS samples appear to be behaving essentially as filled rubbers.

Carbon tetrachloride is a good solvent for both polystyrene and polyisoprene.[4] Electron micrographs for CCl_4 cast SBS show some phase mixing.[4] Phase mixing might also be present in CCl_4 cast SIS reducing the effectiveness of the glassy domains as filler and cross-links and imparting a lower initial modulus than casting from the other solvent systems as shown in Figure 5.

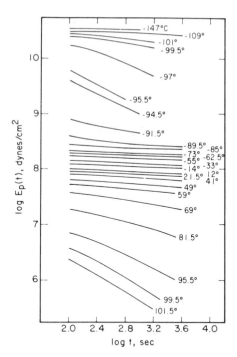

Fig. 6   Stress relaxation isotherms for CCl$_4$ cast SBS.

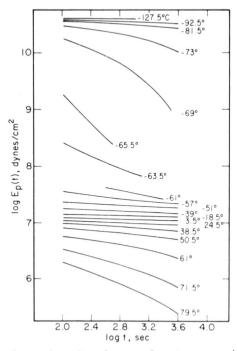

Fig. 7   Stress relaxation isotherms for benzene/heptane cast SIS.

2.   Characteristic Viscoelastic Parameters

Figures 6, 7, and 8 show the reduced modulus versus time at various temperatures for CCl_4 cast SBS, benzene/heptane cast SIS and CCl_4 cast Kraton SIS. The reduced modulus E_p is given by[9]

$$E_p = E_r \left/ \left[\frac{T}{T_o} + \frac{E_r}{E_g} \left(1 - \frac{T}{T_o} \right) \right] \right.$$ (1)

where T = absolute temperature
    T_o = reduction temperature
    E_r = relaxation modulus at T
    E_g = glassy modulus

The reduced modulus isotherms were shifted along the time axis into superposition to give the complete master curves shown in Figures 9 and 10. In general the superpositions were accomplished with little difficulty for both the flexural and tensile data. Also shown in Figure 9 are the master curves for THF/MEK cast[10] and benzene cast[11] SBS. The substantial dependence of the modulus on casting solvent for SBS is quite apparent.

The important feature in Figure 10 is that the effect of solvent is small. The range between the transition and rubbery-flow regions is slightly broader for the benzene/heptane cast sample. perhaps indicating that the phases are purer. This would follow from the contention that casting from CCl_4 leads to some phase mixing or interdispersion which would shift the transition region to longer times and shorten the time at which rubbery flow begins.

Several parameters can be used to characterize viscoelastic behavior.[12] One is the inflection temperature T_i, defined as the temperature at which the relaxation modulus is 10^9 dynes/cm^2 at some fixed time. This fixed time is usually taken as 10 sec, but 100 sec is used here since it is more accurate. Figures 11 and 12 show plots of the 100 sec modulus versus temperature for SBS and SIS. The values for T_i are given in Table 1. Several other parameters are also listed: the glassy modulus E_1; the rubbery-plateau modulus E_2; the negative slope of the master curve in the transition region (n = -d log E_p/d log t); and the negative slope of the modulus-temperature curve in the transition region (s=-d log E_r(100)/dT). Since the modulus is changing rapidly in the rubbery region for the polymers studied here, the E_2's are only approximate intermediate values. For comparison, the characteristic parameters for other copolymers and some homopolymers have also been included in Table 1.

TABLE 1

Characteristic Viscoelastic Parameters of Homopolymers and Copolymers of Butadiene, Styrene, and Isoprene

Polymer	$\log E_1$ (dyn/cm²)	$\log E_2$ (dyn/cm²)	T_i (°C)	n	p	np	s
SBS Block (THF/MEK cast)[10]	10.45	8.87	- 96	.20	.67	.14	.18
SBS Block (benzene cast)[11]	10.57	7.90	- 96	.50	.36	.18	.15
SBS Block (CCl₄ cast)	10.56	8.10	- 93	.72	.49	.35	.28
SIS Block (benzene/heptane cast)	10.60	7.10	- 65	1.33	.39	.52	.51
SIS Block (CCl₄ cast)	10.57	7.05	- 64.5	1.40	.32	.45	.36
Polystyrene[12]	10.30	6.53	93.1	.87	.26	.22	.21
Polybutadiene[12]	10.30	7.03	-106.2	--	--	--	--
SBR (75% butadiene-25% styrene)[12]	10.24	7.44	- 50.8	.71	.34	.24	.16
Polyisoprene[12]	10.24	7.00	- 66.9	1.23	.31	.38	.22
Polyisoprene (vulcanized)[12]	10.39	7.26	- 63.6	1.10	.28	.31	.20

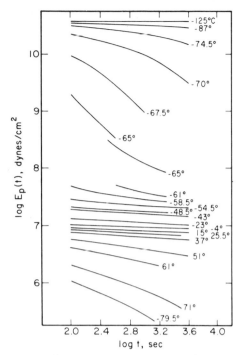

Fig. 8   Stress relaxation isotherms for CCl_4 cast SIS.

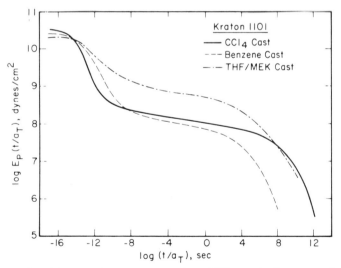

Fig. 9   Viscoelastic master curves for SBS cast from various solvent systems.   Reduction temperature = 25°C.   The data for benzene cast and THF/MEK cast SBS is from Ref. 10 and $E_r(t/a_T)$ is plotted.

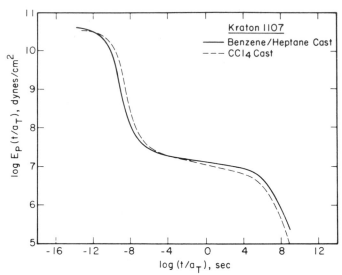

Fig. 10 Viscoelastic master curves for benzene/heptane cast and CCl$_4$ cast SIS. Reduction temperature = 25°C.

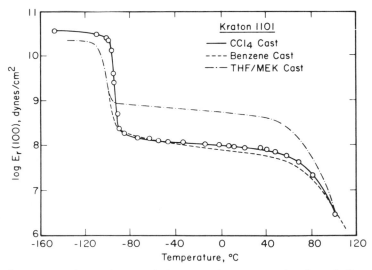

Fig. 11 Stress relaxation modulus at 100 seconds for SBS cast from various solvent systems. Data for benzene cast and THF/MEK cast SBS is from Ref. 9.

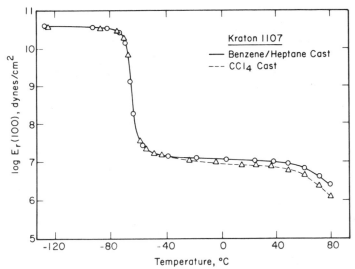

Fig. 12   Stress relaxation modulus at 100 seconds for benzene/
heptane cast and CCl_4 cast SIS.

Looking first at SBS, casting from CCl_4 gives no significant
change in E_1; however, E_2 falls between the widely separated values
for samples cast from THF/MEK and benzene.   This last result agrees
with the findings of Beecher et al.[4]   Also notice that the CCl_4 cast
sample has a T_i 3°C higher than the others.   This could be due to
phase mixing which would raise the glass transition temperature of
the polybutadiene phase with which T_i is closely associated.   All
the SBS samples have a rubbery modulus greater than that for either
polybutadiene or SBR (styrene-butadiene random copolymer) showing
the effect of the polystyrene domains.

Another parameter for SBS which is greatly affected by the
casting solvent is n.   The CCl_4 cast sample has the higher value,
equal to that for SBR, and consequently has the steepest and narrow-
est transition region.   Interphase mixing probably plays a role here
also.   It has been found[13] for unvulcanized natural rubber (polyisoprene) that the addition of fine carbon black filler tends to
broaden the transition region, viz., decrease n.   Crosslinking has
a similar effect, e.g., compare vulcanized and unvulcanized polyisoprene in Table 1.   One might then expect n to increase as the
effectiveness of the polystyrene domains as filler and crosslinks
decreases.   This seems to be the case.   The domains are least
effective in the CCl_4 cast sample where there is some phase mixing,
more effective in the benzene cast sample where the phases are
relatively pure, and most effective in the THF/MEK cast sample
where there are numerous interconnections among the domains.   The
filler aspect of the argument, however, does not strictly hold for
the THF/MEK cast sample because the polystyrene phase is continuous.

Shen et al.[10] have suggested that n may reflect the number of transitions in an interfacial region, a higher value of n indicating fewer transitions. This hypothesis would again predict CCl_4 cast Kraton 1101 to have the largest n since some phase mixing would reduce the compositional difference between the polybutadiene and polystyrene regions and consequently the number of transitions.

Turning now to SIS, the values of E_1, E_2, n, and T_i for the two methods of preparation are nearly the same, indicating very little difference in morphology. The fact that E_2 is lower and T_i and n slightly higher for the CCl_4 cast sample is further evidence that the phases are less pure than in the benzene/heptane cast sample. The T_i's for both samples are higher than that for polyisoprene as would be expected because of the polystyrene domains. The difference between n's can be explained in terms of a filler effect and an interfacial region as was done for SBS. One interesting finding is that there is only a small increase in E_2 for SIS over that for polyisoprene. For other SIS copolymers, increases of almost an order of magnitude have been reported.[13]

3.   Time-Temperature Shift Factors

The validity of Time-Temperature superposition principle to obtain viscoelastic relaxation master curves has already been established for block copolymers.[1-3] However, the shift factors do not follow the classical Williams-Landel-Ferry equation.[15] This is not surprising since WLF theory is based on a single reference or glass transition temperature which is clearly not the case when multiple phases are present. These data are shown in Figures 13, 14, and 15.

Shen and Kaelble[10,11] recently found that shift factors obtained from stress relaxation data on benzene cast SBS and a plasticized SBS (Thermolastic 125) exhibit three distinct regions of temperature dependence. In the vicinities of the polybutadiene and polystyrene glass transitions, the shift factors a_T are well defined by the following special form of the WLF equation with "universal" constants:[15]

$$\log a_T = \frac{-16.14 \left(T - T_d \right)}{56 + T - T_d} \qquad (2)$$

where T_d represents a characteristic temperature close to the glass transition temperature (T_g). Intervening the low and high temperature regions of WLF behavior is a broad transition region of constant

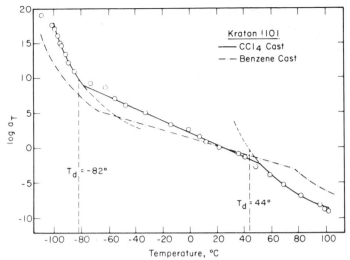

Fig. 13   Shift factors for CCl$_4$ and benzene cast SBS.   Reference temperature=25°C.   Data for benzene cast SBS is from Ref. 10.

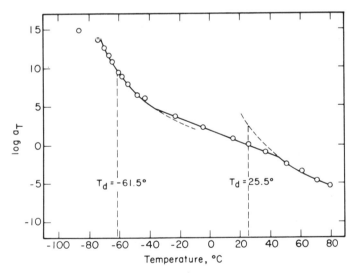

Fig. 14   Shift factors for benzene/heptane cast SIS.   Reference temperature=25°C.

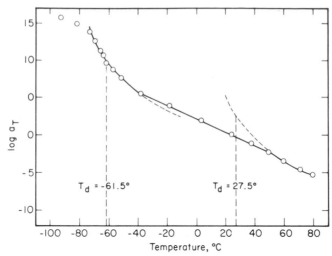

Fig. 15   Shift factors for CCl$_4$ cast SIS.   Reference temperature
=25°C.

slope on the log a_T versus T plot.   Equation 2 can still be applied
in this range, however, if T_d is allowed to change, i.e., each
point is on a different WLF curve.   The constant slope implies
that each point is from the same relative position on its respective WLF curve.   Thus $T-T_d$ must be a constant.*   Shen and Kaelble
believe that in this intermediate region the molecular mechanisms
for stress relaxation are being contributed by an interfacial
phase that intervenes the polybutadiene continuum and the spherical
polystyrene domains.   They interpret this interfacial phase as
being a series of spherical shells enclosing each of the pure
polystyrene domains and characterized by a fairly sharp concentration gradient between shells of mixed polybutadiene and polystyrene.   The outermost shell has the lowest T_d and the lowest
concentration of polystyrene segments.   Thus the condition of
constant $T-T_d$ indicates that as the temperature is raised in the
linear region, stress relaxation occurs in succeeding interior
shells of higher $T_d \approx T_g$.   The presence of a linear region has also
been reported by other workers.[4,16]

---

*Equation 2 can not be used to reproduce an entire shift factor
plot because it must be recognized that when T_d changes a new reference temperature is being defined.   Thus, the condition $T-T_d$
does not imply that log a_T is independent of temperature.

This modified WLF theory is applied to CCl_4 cast SBS in Figure 13. It is seen that the classical WLF equation (Eq. 2) can be applied in its usual form between -80°C and -100°C with T_d=-82°C and about 50°C with T_d=44°C, where T_d has been selected to give the best fit. Below -100°C the polymer is in the glassy state where WLF theory does not apply, and consequently the shift factors diverge. The broad intermediate range from -80° to 50°C is approximately a straight line, although there is some scatter near the transitions to WLF behavior. This linear region can be described in Equation 2 with $T-T_d$=45.7°C. The slope is -.0874(°C)$^{-1}$ compared to -.0578(°C)$^{-1}$ for benzene cast[10] and -.116(°C)$^{-1}$ for compression molded[16] SBS. The shift factors for benzene cast SBS are also shown in Figure 13. As can be seen, the linear region is narrower for the CCl_4 cast sample. This can be interpreted to mean that casting from CCl_4 results in fewer interfacial transitions because of phase mixing.

The shift factors for both benzene/heptane and CCl_4 cast SIS are also described very well by the Shen-Kaelble model as shown in Figures 14 and 15. Again the diverging points at the lowest temperatures are in the glassy region where WLF theory does not apply. The various fitting parameters are summarized in Table 2.

The shift factors yield another viscoelastic parameter, namely, the negative slope p of log a_T versus T at T_i. According to Tobolsky,[12] s=np. This relation is obeyed well by benzene/heptane cast SIS and to within 25% by CCl_4 cast SIS and SBS as seen in Table 1.

4. Comparison of Viscoelastic and Dielectric Behavior

The preceding data clearly support the existence of an interfacial region between the pure polystyrene and pure polybutadiene or polyisoprene domains. These observations are in consonance with recent theoretical predictions.[17,18] However, other experimental investigations[4,19] have shown that certain interfacial regions appear to exhibit a primary glass transition. It was observed as a peak in the mechanical loss tangent data. In order to examine this phenomenon, dielectric measurements were taken of Kraton 1101 cast from carbon tetrachloride, benzene/heptane and methylethylketone/tetrahydrofuran. Figure 16 shows the dielectric loss data of the CCl_4 cast Kraton 1101. The peaks are those of the primary glass transitions of the polymer blocks. There is no sign of a peak in the intermediate region that may be attributed to the interfacial phase. Data for the sample cast from the other two solvents are similar, and thus will not be reproduced here.

TABLE 2

Shift Factor Parameters for Triblock Copolymers
Described by the Shen-Kaelble Model

Polymer	Low Temperature T_d(WLF), °C	Slope	Linear Region $T-T_d$, °C	Range, °C	High Temperature T_d(WLF), °C
SBS (CCl$_4$ cast)	-82	-.0874	45.7	-80 to 50	44
SBS (benzene cast)[11]	-96	-.0578	66	-50 to 80	60
SIS (benzene/heptane cast)	-61.5	-.0914	43.5	-40 to 50	27.5
SIS (CCl$_4$ cast)	-61.5	-.0752	53.7	-37 to 45	25.5

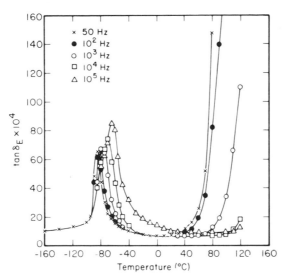

Fig. 16 Dielectric loss tangent for CCl_4 cast SBS as a function of temperature.

        In order to compare with the dielectric loss data, the mechanical loss curve for CCl_4 cast Kraton 1101 is computed from its viscoelastic master curve (Figure 9). Calculations were performed by the Linear Programming technique on a CDC 6400 computer. Detailed description of this technique has been given in another publication[20] and will not be repeated here. Figure 17 shows the calculated loss tangent data for the CCl_4 cast Kraton 1101. The two maxima may be associated with the primary glass transitions of polybutadiene and polystyrene. Again no evidence of an intermediate transition is discernible. It is possible that the experimental techniques used here are not sensitive to the type of transitions in question. However, these observations are nevertheless consistent with the Shen-Kaelble model, which assumes a distribution of mixed phases. One would only expect an observable glass transition if the mixed region is a homogeneous one. It is possible that the apparent discrepancy between our data and those reported in the literature[4,19] is due to differences in sample preparation. It would be interesting to speculate on the likelihood of observing a "triple-WLF" type of viscoelastic shift factor data for a block copolymer consisting of two pure phases and a homogeneous interfacial phase.

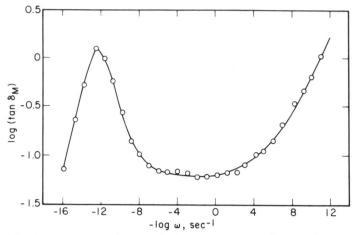

Fig. 17 Dynamic mechanical loss tangent for CCl_4 cast SBS calculated from stress relaxation master curve by Linear Programming. (Reference temperature=25°C)

ACKNOWLEDGEMENT

The authors wish to thank Messrs. A. Kaya and A. Ling for assisting in some of the experiments. Partial support by the Petroleum Research Foundation, administered by the American Chemical Society, is gratefully acknowledged.

REFERENCES

1. J. Moacanin, G. Holden and N. W. Tschoegl, eds., "Block Copolymers," Interscience, New York, 1969.

2. S. L. Aggarwal, ed., "Block Polymers," Plenum, New York, 1970.

3. G. E. Molau, ed., "Colloid and Morphological Behavior of Block and Graft Copolymers," Plenum, New York, 1971.

4. J. F. Beecher, L. Marker, R. D. Bradford and S. L. Aggarwal, J. Polymer Sci., C, 26, 163 (1969).

5. J. L. Binder, Anal. Chem., 26, 1877 (1954).

6. H. Hendus, K. H. Illers and E. Ropte, Kolloid Z. u. Z. f. Polym., <u>216</u>, 110 (1967).

7. A. C. Ouano, J. Polymer Sci., Part A-1, <u>9</u>, 2179 (1971).

8. T. Uchida, T. Soen, T. Inoue and H. Kawai, J. Polymer Sci., A-2, <u>10</u>, 101 (1972).

9. E. Catsiff and A. V. Tobolsky, J. Colloid Sci., <u>10</u>, 375 (1955).

10. M. Shen, E. H. Cirlin and D. H. Kaelble, in G. E. Molau, op. cit., p. 307.

11. M. Shen and D. H. Kaelble, J. Polymer Sci., B, <u>8</u>, 149 (1970).

12. A. V. Tobolsky, "Properties and Structure of Polymers," Wiley, New York, 1960.

13. A. R. Payne, in P. Mason and N. Wooley, eds., "Rheology of Elastomers," Pergamon, London, 1958, p. 86.

14. J. F. Henderson, K. H. Grundy and C. Fisher, J. Polymer Sci., C, <u>16</u>, 3121 (1968).

15. J. D. Ferry, "Viscoelastic Properties of Polymers," 2nd ed., Wiley, New York, 1970.

16. T. L. Smith and R. A. Dickie, J. Polymer Sci., C, <u>26</u>, 163 (1969).

17. D. F. Leary and M. C. Williams, J. Polymer Sci., B, <u>8</u>, 335 (1970).

18. D. H. Kaelble, Trans. Soc. of Rheo., <u>15</u>, 235 (1971).

19. T. Miyamoto, K. Kodama and K. Shibayama, J. Polymer Sci., A-2, <u>8</u>, 2095 (1970).

20. M. Shen, R. T. Jamieson and M. Drandell, J. Polymer Sci., <u>C35</u>, 23 (1971).

DYNAMIC MECHANICAL PROPERTIES AND STRUCTURE OF WHITE ASH (FRAXINUS
AMERICANA L.) WOOD

D.E. Kline, R.P. Kreahling and P.R. Blankenhorn

Departments of Material Sciences and Wood Sciences

The Pennsylvania State University

ABSTRACT

Wood, a natural polymer, is characterized by complex dynamic
mechanical properties which are a function of temperature, specimen history and other factors. For white ash (Fraxinus americana
L.), a relaxation process, believed to be associated with presence
of water, is present in the $200^{\circ}K$ region. Damping peaks are also
present near $360^{\circ}K$ and $550^{\circ}K$, and these give evidence of other
relaxation mechanisms in wood. The relaxation behavior below room
temperature has been studied with regard to both water and another
additive (ethylenediamine). DSC scans and measurements of effluent
released as a function of temperature illustrate the changes taking
place in the material during testing. Carbonization of wood in a
N_2 atmosphere has been studied and the results obtained are discussed. Scanning electron micrographs, taken on the carbonized
specimens, illustrate the effects of the treatment and the ultrastructure of the ash system.

INTRODUCTION

Background of Wood Studies

By far, the largest crop harvested by man is wood. This complex material is a natural polymer composite consisting essentially
of cellulose fibers, hemicellulose and lignin. Wood has been used
by man for shelter, tools, charcoal to make iron, paper, furniture,
toys, and nearly a limitless number of other items. Today in the

United States, consumption amounts to tons per year per capita. Paper products alone constitute about 600 pounds per capita and thus a rising population each year places increasing demands for more raw material. The industry related to wood amounts to tens of billions of dollars.

Contrary to possible first thoughts, relatively little is known about many aspects of wood structure, behavior, and properties. When wood is used as part of a composite system or when it undergoes a particular treatment, the result often seems to be unpredictable. It is an intriguing and challenging material for an endless number of scientific studies. At the same time, it is a substance where increased knowledge can lead to many practical benefits.

For many reasons the wood products industry is generally somewhat less prosperous than might be considered satisfactory. Certainly the time is coming when it will be vital to know far more about wood and to learn how to use economically the enormous quantities of this material that are now largely wasted. Already the problems associated with storing or burning substances, such as unwanted wood bark, as well as the problem of pollution of the streams by waste pulping liquor containing large quantities of lignin, bring to focus the great need for more knowledge concerning wood utilization and wood properties.

The results presented here are concerned with the dynamic mechanical behavior of ash wood and other species, especially with regard to the effects of moisture content[*] and of other additives. The role of water in wood is, of course, an old problem of considerable importance, and these studies may help to clarify some of its effects. Various heat treatments have been carried out with the object of delineating the changes that occur in both material properties, as for example the dynamic mechanical properties (DMP), and structure. With rate-controlled heat treatment under a nitrogen atmosphere (rate-carbonized), many of the fine structural features of the wood can be retained. Scanning electron micrographs made on such samples reveal much interesting detail.

Literature Concerning Dynamic Testing of Wood

Some years ago an interesting study by Skudrzyk revealed that important musical properties of some wooden instruments were

----

[*] Moisture content (m.c.) is usually expressed in weight percent. It is given by the weight of the wood with moisture minus the oven-dry weight of the wood divided by the oven-dry weight of the wood times 100.

detectable from dynamic mechanical tests. His results showed that
certain types of aged wood were necessary to obtain the sound of
the excellent violins, such as the Stradivarius violin. The tone
is related to the damping characteristics of the wood as a function of frequency and thus to the relative damping rate of harmonic
tones. The vibrational properties of some woods used in musical
instruments have also been studied by Fukada (8).

Hearmon has discussed the theory of vibration testing of wood
and the influence of shear stress and rotatory inertia on the free
flexural vibration of wooden beams (4, 5). The effect of frequency
on the DMP of wood (at room temperature) has been studied by several
investigators (6-9). Relations between dynamic and static modulus
and between creep and static modulus have been reported (10, 11,
13, 14). Jayne (9) used DMP analysis to study glue line phenomena
in wood, and Pillar (12) showed that it can be a sensitive method
for investigating curing properties of resorcinol resin in contact
with wood. At room temperature, a relation between moisture content, moisture gradients and internal friction was reported by
Moslemi (15). Pentoney (6) noted that the internal friction (audio
frequencies) was at a minimum value at about 6% moisture content
and then increased with increasing moisture content (m.c.) up to
the fiber saturation point*.

Studies of the dynamic mechanical properties over a wide temperature range have not been extensive. For oven-dry wood, Fukada (8)
reported data from 245 to 375°K. Bernier and Kline (2) investigated
both birch and PMMA-impregnated birch from 90 to 475°K and they
demonstrated that both moisture and PMMA impregnation had a pronounced effect on the DMP spectrum. At low temperatues (< 200°K),
increased moisture resulted in an increase in modulus. James (3)
measured the DMP behavior for douglas fir at moisture contents of
1.8 to 27.2% from 255 to 366°K. At the lower temperatures (∿ 255°K),
there appeared to be an internal friction maximum for douglas fir
samples containing less than 7.2% m.c., and for samples containing
greater than 12.8% m.c., the onset of another damping maximum
appeared near 360°K.

Current and past studies (16) of white ash and other species
indicate that water, and many other additives, can alter the complex DMP spectrum of wood, often in a systemic manner. (A detailed
study emphasizing the role of water in altering the DMP of black
cherry [17] will be reported soon.) Furthermore, high temperature
DMP testing of wood necessarily alters the material drastically,

---

*The fiber saturation point is that point when all water is
absent from the cell lumens, but the cell walls are fully saturated
with water.

and this, in part, has been responsible for investigations of the
effect of heat treatment on the DMP behavior and on structure.
This paper will discuss some of the dynamic mechanical properties
of white ash (Fraxinus americana L.) from $100^{\circ}K$ to $> 500^{\circ}K$, the
effects of moisture and of ethylenediamine in the 100 to $300^{\circ}K$
region, the DMP behavior following a carbonization treatment, and
some results of a scanning electron micrograph study of the carbonized structure.

SPECIMENS AND PROCEDURES

Specimens

Characteristics of specimens used are given in Table 1. All
of these samples were obtained from a white ash tree in Central
Pennsylvania which was cut in 1966. When the logs were brought to
the mill and cut, sample material was obtained and sealed in a
polymer film to be kept in a freezer for storage.

From the above stock, specimens were machined to rods 0.64 cm
in diameter and 11 cm in length, resealed in a polymer film and
stored in a freezer until tested. Specimens had their long axis
parallel to the longitudinal direction of the tree. The specimens
contained no noticeable defects. Before testing, some of the
specimens were conditioned in a temperature and humidity chamber
to bring the moisture content to a given level. These levels were
actually estimated before the test, then confirmed afterward using
the oven-dry weight. It has been found that tests carried out
between the temperatures from 100 to $275^{\circ}K$ result in a change in
moisture content of less than 0.5%; but, above room temperature,
the moisture content of the wood samples changes rapidly during
testing.

The rate-carbonized specimen mentioned in Table 1 received a
heat treatment before testing in a nitrogen atmosphere in which
the temperature was gradually increased, at a rate of about $3^{\circ}K/$
minute, to a given temperature and then held at this temperature
for two hours. They were permitted to remain in the oven during
slow cooling.

Procedures

The density of the specimens was determined by mass-dimension
measurements. Dynamic mechanical tests were carried out by an
apparatus which is a modification of that developed by Kline (18).
Frequencies were typically in the range of about 800 to 2500 Hz and
the testing temperatures ranged from near $100^{\circ}K$ to as high as $600^{\circ}K$.

TABLE 1.   CHARACTERISTICS OF SPECIMENS – WHITE ASH (FRAXINUS AMERICANA L.)

Specimen	m.c.%	Mass (gm)	Length (cm)	Diam. (cm)	Q^{-1} Peak Temp. °K	Resonant Freq. (Hz) 100°K	275°K	E' x 10^{-10} dynes/cm^2 100°K	275°K
As-cut		2.99	11.17	.64		2251	1831	20.9	13.81
		1.57	11.07	.56					
After									
1	23.4	2.513	10.95	.63	185	2121	1727	15.5	10.2
4	7.2	2.366	10.76	.59	217	2452	2180	23.2	78.3
3	11.6	2.226	10.71	.61	201	2245	1910	15.2	11.4
2	15.8	2.501	10.85	.62	195	2387	2010	19.6	13.9
2 (second run)	0.4	2.168	10.84	.59	225	2294	2113	18.1	15.3

5 (rate-carbonized ash)*

Oven-dry Condition	After Carbonization
ℓ = 10.78 cm	ℓ = 8.95 cm
d = .59 cm	d = .44 cm
w_{od} = 2.3102 gm	w = 0.7116 gm

	Resonant Frequency (Hz) 100°K	300°K	E' x 10^{-10} dynes/cm^2 100°K	300°K
	1447.3	1452.7	4.78	4.81

*Rate-carbonized to 600°C at a rate of 3°K/minute and held for 2 hours.

In the Kline apparatus, specimens are supported by vertical cotton threads in a horizontal position and driven in a transverse mode. The elastic storage modulus, E', at each temperature is calculated using the resonant frequency, f_o, of the first transverse mode, the density, ρ (gm/cm^3), and the dimensions of the specimen. One form of the equation is $E' = \ell^4 \rho \, f_o^2/12.7 \, R^2$ where ℓ(cm) is the specimen length and R(cm) is the radius of gyration of the specimen cross section. Room temperature parameters were used in obtaining the elastic modulus, and thus the values of E' at other temperatures are only nominal. It has been shown from calculations on various polymers that nominal values are usually within a few percent of true values.

The internal friction, Q^{-1}, is calculated from the half-width Δf of the resonance curve and the resonance frequency; i.e., $Q^{-1} = \Delta f/f_o$. From this, other estimates of the damping capacity can be readily calculated. In polymer systems, and especially in wood, where water is an important parameter, it should be noted that the material properties can be altered (by changes of temperature and testing environment) during the several hours required to complete the testing. The temperature of the specimen is typically increased at a constant rate of about 1 K^o/minute during an experimental run. Slowly flowing nitrogen gas is used as the atmosphere for the specimen.

Scanning electron micrographs of rate-carbonized specimens were obtained on a Type-JSM, Japan Electron Optics Laboratory Company, Ltd., scanning electron microscope (SEM). It has been found that carbonization of wood under the conditions described leaves the ultrastructure very nearly intact and results in a material which is sufficiently conductive to be directly useable in the SEM apparatus (without coating). Detail and depth of field are excellent for these conditions.

RESULTS AND DISCUSSION

Damping data and modulus data, as a function of temperature, are presented in Figs. 1, 2 and 3 for the specimens of white ash (Fraxinus americana L.) of varying moisture content and also for the carbonized sample. Note that there is a large damping peak in the vicinity of 200°K and that this peak is suppressed with decreasing moisture content. These results are generally consistent with results obtained on other wood species and also with results obtained for polymers, particularly those where water is sorbed.

Bernier and Kline (2) have presented results for birch samples (6% m.c.) which indicated an unsymmetrical internal friction peak

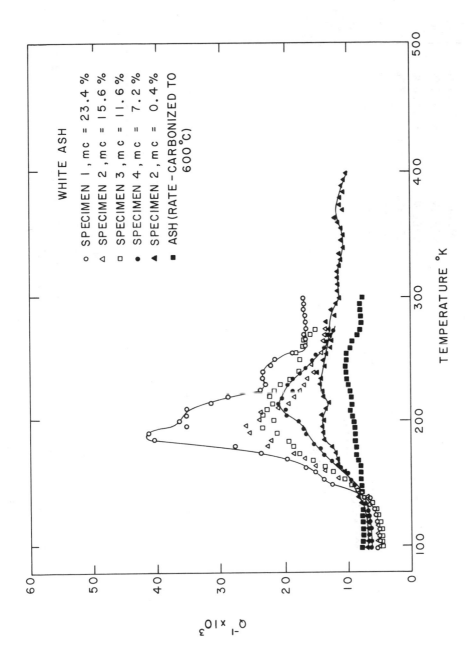

FIG. 1, INTERNAL FRICTION OF WHITE ASH AND RATE - CARBONIZED WHITE ASH AS A FUNCTION
OF TEMPERATURE AND MOISTURE CONTENT.

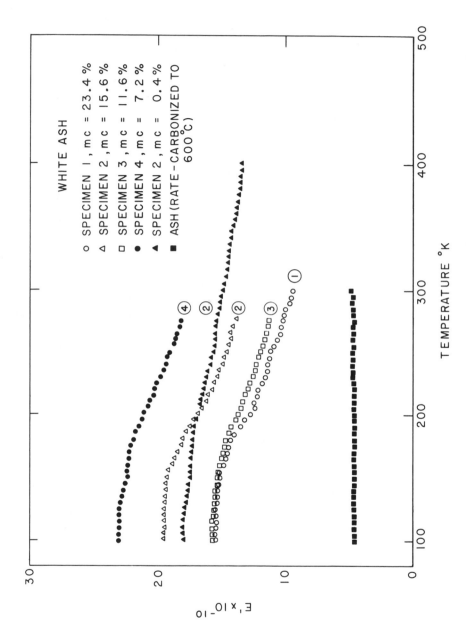

FIG. 2, DYNAMIC ELASTIC MODULUS OF WHITE ASH AND RATE–CARBONIZED WHITE ASH AS A
FUNCTION OF TEMPERATURE AND MOISTURE CONTENT.

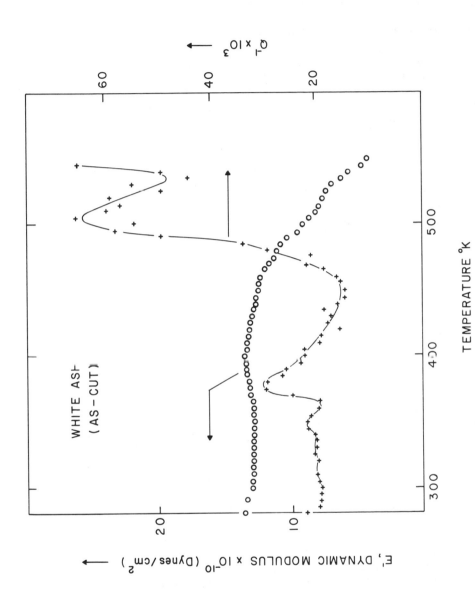

FIG. 3 , DYNAMIC ELASTIC MODULUS AND INTERNAL FRICTION OF WHITE ASH FROM 300 TO 550°K.

near 215°K, a relaxation in the 360°K region, and an onset of a
large relaxation approaching 500°K. Compared with an oven-dried
specimen ($\sim$ 0% m.c.) they showed that the low temperature peak
changed shape and temperature position.

The 360°K peak was also observed to change with changing
moisture content (2, 17). In investigating the relaxation near
360°K as a function of temperature and moisture content, it was
recognized that above 273°K, and especially above room temperature,
the moisture content of the wood changed as the test proceeded.
Also, at higher temperatures (> 373°K) important irreversible
changes progressively took place in the wood (16). These changes
are known to alter the mechanical relaxation in the neighborhood
of 200°K (16).

Specific Effects of Water Below Room Temperature

The data of Figure 1 show that a large low temperature, Q^{-1},
peak occurs for specimen 1 (23.4% m.c.) near 185°K with possible
shoulders above 200°K. These effects are probably associated with
water in the cell wall. Somewhat similar results for high water
contents in dielectric studies of Poly (2,6-dimethyl phenylene
oxide) were also noted (22). The data for the other specimens
containing less water indicate that, as the moisture content
decreases, the total area under the relaxation peak tends to
decrease, and the temperature (at which the main maximum occurs)
increases. This effect has also been noted by Allen, et al (28)
and others (29) in several different polymers. At very low
moisture content (specimen 2 rerun, 0.4% m.c.), the Q^{-1} values
increase only gradually with increasing temperature to form a low,
very broad peak over the temperature range from about 180 to 280°K.

The dynamic modulus, E', data of Fig. 2 are consistent with
the Q^{-1} data in that, with increasing temperature, most samples
exhibit inflection regions near 200°K, in addition to a general
overall decrease in value with increasing temperature possibly
due to lattice expansion effects. The overall E' levels vary
from specimen to specimen. This may be due to variability of
wood moduli taken from different sections of the same tree.
However, it is important to take note of the E' data for specimen 2 which was first run at 15.6% m.c. and then run again at
0.4% m.c. (same specimen, second run). Below $\sim$ 200°K the higher
water content resulted in the wood being more rigid while at
higher temperatures the water produced a plasticizing effect in
that modulus values are lower.

The relaxation behavior of the wood at low temperatures
appears to be a combined result of several processes. For

specimen 2 rerun (0.4% m.c.) the Q^{-1} data show that a broad relaxation is still present even though nearly all of the water has been removed. It is likely that the remaining water is hydrogen bonded in the cell wall. About 42% of the wood is considered to consist of cellulose. The observed relaxation is probably related, as in synthetic polymers also (20), to motions of hydrogen bonded water molecules, or water-polymer chain complexes, in regions of limited order.

In Fig. 4 the temperature position of the principal Q^{-1} maximum is plotted as a function of moisture content for three different woods; viz., white ash, black cherry, and hard maple (17). For moisture contents up to ∿ 6%, the peak position is about 225°K. Up to about 6% m.c. the H_2O molecules tend to be hydrogen bonded in a monomolecular layer in the structure (20), and it appears probable that the Q^{-1} peak for specimens up to ∿ 6% m.c. may be associated with hindered motion of the water units* associated with the primary or secondary OH side groups located in the disordered regions. Results for polymers such as nylon (1) indicate that water plays a somewhat similar role only here bonding is to the amide groups in the amorphous regions.

In amounts greater than about 6% m.c. the water in wood tends to cluster in the structure and solidify at low temperatures. It is known from other work on polymers (1, 29) that free or bound water attains enough internal mobility to give rise to internal friction effects at temperatures well below 273°K. This mobility usually occurs at temperatures above 180°K. In wood at slightly higher temperatures and NMR frequencies mobility has been indicated (21). This may explain the increase in the overall area of the Q^{-1} peak with higher moisture content. Note also that for samples 1, 2 and 3 of Fig. 1 the data indicate an overlapping of relaxations, and this is to be expected if water is present in different forms. There are indications in Fig. 1 that the solidified water in wood begins to attain some mobility, sufficient to contribute to the internal friction, at temperatures as low as 150°K. It seems clear from the above that the composite damping peaks in the wood samples containing moisture are thus a result of the contributions of several mechanisms and are related to the manner in which the water molecules are bonded.

---

*Although wood swells with increasing moisture content, X-ray studies indicate that unit cell dimensions of the cellulose in wood do not change with water content (20). The very well ordered cellulose apparently resists much penetration of the water, but many other sites for hydrogen bonding are available in the cellulose, hemicellulose and other constituents.

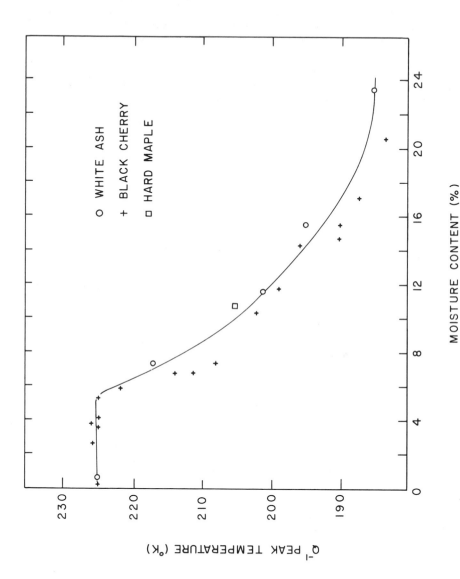

FIG. 4, INTERNAL FRICTION PEAK TEMPERATURE VS MOISTURE CONTENT FOR WHITE ASH, BLACK CHERRY AND MAPLE.

Several different species of wood have now been tested and all results to date show low temperature relaxation peaks, regardless of the species or the portion of the tree where the specimen was cut. In instances where the effects of moisture content and temperature were examined in detail, the results are similar to those presented here for ash, and a more detailed account of these studies will be reported later (16, 17).

The overall decrease of E' with temperature for most specimens is probably associated with an expansion of the structure with temperature accompanied by a decrease in intermolecular forces. For specimens with appreciable moisture content, the increase in E' with decreasing temperature below 200°K can be attributed to solid H_2O acting much like a filler material, with the hindrance of a steric or geometric nature. Above 200°K, the water molecules acquire mobility and thus no longer act as rigid filler particles, and the modulus drops. This type of effect, observed in polyimides (22, 27) and other polymers (1), has recently been attributed to similar phenomena (22).

Effect of Ethylenediamine on the DMP

Figure 5 presents DMP results for a wood specimen treated with ethylenediamine (EDA). The treated specimen was first dried at room temperature (RT), soaked in H_2O and tested to 400°K resulting in a density (after test) of about 0.738 gm/cm^3, a mass of 2.424 gm, a length of 11.35 cm, and a diameter of 0.61 cm. Upon subsequently soaking this specimen for 210 hours (RT) in EDA, it attained a density of 1.151 gm/cm^3, a mass of 4.523 gm, a length of 11.37 cm, and a diameter of 0.66 cm.

The plasticizing effect of the EDA is evident in the modulus and damping results of Fig. 5. As in the case of water in ash, the effect of the EDA is to make the lattice more rigid at low temperatures, indicating that it is acting in the manner of a rigid filler particle. Near 195°K, the E' curve of the treated specimen drops rapidly as the EDA begins to act as a plasticizer.

Accompanying the large inflection in E' in this region is a large relaxation peak (see also Fig. 5) with a slight hump near 150°K. The main relaxation is so strong that some of the data for the maximum cannot be obtained. This is a result of the high EDA content of this specimen, as the strength of the relaxation is probably related directly to the amount of EDA present.

Use of steam to plasticize wood is a well-known technique. More recently, anhydrous ammonia has been used with good results (23), but the practical problems of handling ammonia are severe. In tests

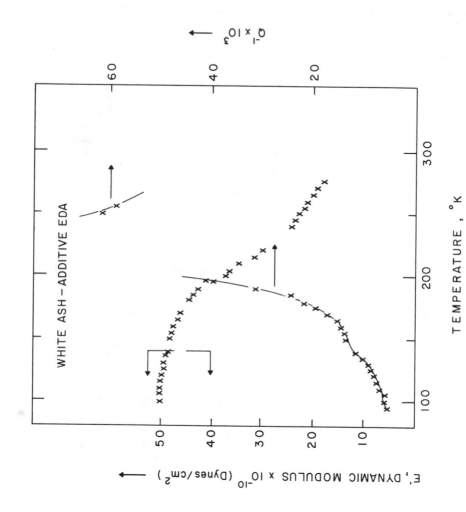

FIG. 5, DYNAMIC ELASTIC MODULUS AND INTERNAL FRICTION OF WHITE ASH AND ASH TREATED WITH ETHYLENEDIAMINE AS A FUNCTION OF TEMPERATURE.

with EDA and other amines efforts are being made to achieve useful
and practical results for bending and forming wood materials by
plasticization. This could lead to better fabrication, less waste,
and overall better wood utilization.

DMP Effects Above Room Temperature

Modulus and damping data are presented in Fig. 3 for an as-cut
white ash sample (starting ρ = 0.845 gm/cm^3). A Q^{-1} peak is present
near 375°K, a large peak appears in the 500°K region, and there may
be multiple peak structure, possibly associated with thermal degradation of the wood structure (24, 25). The E' data show inflection-
like behavior in these regions; and in addition, a very rapid
decline near 550°K. It is important to note that the density and
dimensions probably change with time during the test, but the modulus
data presented are based on measurements taken before placing the
specimen in the test chamber. Thus, the overall E' level is in
error, and the unusual rise of E' with temperature near 400°K may
not be real although the modulation on the curve probably has
meaning. The results presented here are consistent with earlier
DMP results (2, 16, 17).

Figure 6 presents data obtained from differential scanning
calorimetry (DSC) runs and from measurements of effluent released.
Tests were made on a piece of the as-cut specimen as a function of
temperature. Internal friction measurements were also made in a
similar sample and these data are also given in Fig. 6. The rate
of temperature scan is higher but the sample size is smaller in
the DSC-effluent tests as compared to the DMP tests. Effluent in
the wood specimens in this temperature range is mostly water and
this begins to come off rapidly above 320°K. Note that internal
friction and DSC peaks are present in the region above 320°K. The
375°K peak in the Q^{-1} data (Fig. 3) is probably related to some
internal changes accompanying loss of water. For specimens < 6% m.c.
the damping peak at 350°K to 400°K is nearly absent (17). The
modulus (E') drops rapidly as the temperature approaches 550°K, the
rate of effluent discharge again rises rapidly, and the DSC scan
goes through another maximum. These effects are an indication of
severe degradation of the wood structure (24).

Results for Rate-Carbonized Wood

It has been found that rate-carbonization of wood in a nitrogen
atmosphere can lead to interesting results. The usual procedure
is to raise the specimen temperature gradually, in flowing nitrogen,
at a rate of about 3°K/minute and then hold at the maximum temperature for two hours. For specimens treated to temperatures of about

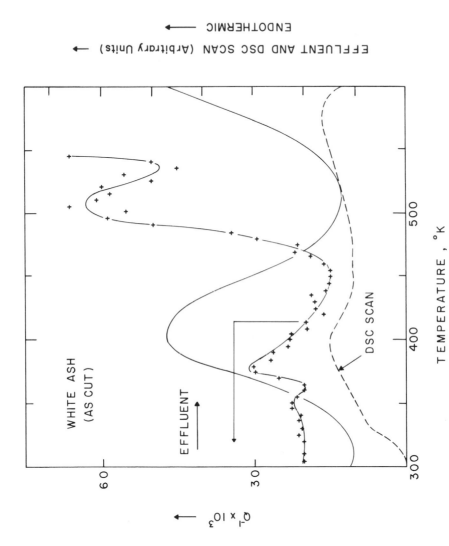

FIG. 6, INTERNAL FRICTION, EFFLUENT AND DSC SCAN OF WHITE ASH AS A FUNCTION
OF TEMPERATURE.

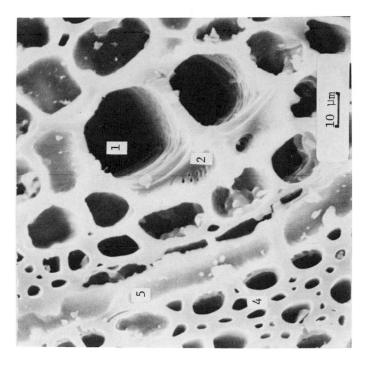

Fig. 8. Scanning electron micrograph of ash carbonized to 873°K. This is a cross-sectional view of a (1) late wood vessel with: (2)pitting (3) longitudinal parenchyma cells (4) fibers and (5) ray parenchyma cells.

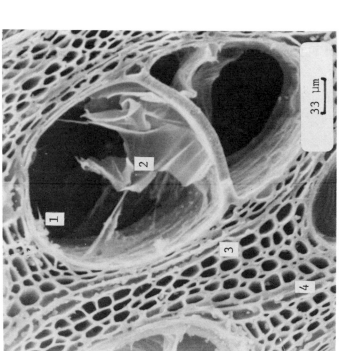

Fig. 7 Scanning electron micrograph of ash carbonized to 873°K. This is a cross-sectional view of an (1) early wood vessel. Clearly evident are: (2) Tyloses (3) fibers and (4) ray parenchyma cells.

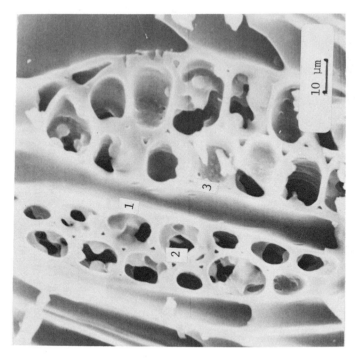

Fig. 10 Scanning electron micrograph of ash carbonized to 873°K. This is a tangential view of (1) ray parenchyma cells showing (2) intercellular spaces and (3) end wall pitting.

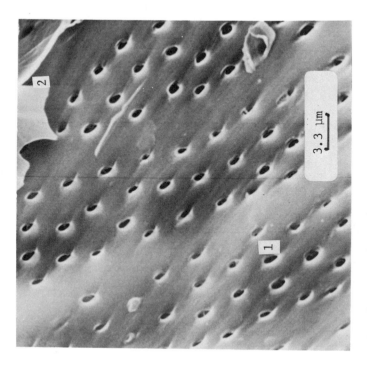

Fig. 9 Scanning electron micrograph of ash carbonized to 873°K. This is a radial view of a vessel displaying (1) pitting and (2) tyloses. This vessel is adjacent to a ray cross field.

$600^{O}K$, subsequent DMP tests show that the elastic modulus decreases compared to that of normal wood, but heat treatment to higher temperatures than $600^{O}K$ causes the E' values to reverse and to increase with increasing heat-treatment temperature (26).

From Fig. 1, the Q^{-1} values for a specimen treated to $873^{O}K$ are observed to be relatively temperature independent with an indication of only a slight peak near $250^{O}K$. At low temperatures ($100^{O}K$) the Q^{-1} values are higher than for uncarbonized wood. The E' data (Fig. 2) are almost temperature independent with some very slight modulation, and there is no indication of any significant relaxation. The overall E' level is very low compared to uncarbonized white ash; however, as compared to many types of wood chars, the bending strength appears to be relatively high as determined from casual observation. These DMP changes with heat treatment are being investigated further. The general behavior is similar to that observed for vitreous carbon in earlier studies (19).

Another interesting effect of the heat-treating procedure is that the wood ultrastructure appears to remain nearly intact as noted by comparing it to uncarbonized material; and, since the carbonization makes the material more electrically conductive, the surface structure can be observed directly in the scanning electron microscope without use of a coating. Because the depth of field is large and the contrast is excellent in this system, the carbon skeleton of the wood reveals many interesting details of the wood structure.

Figures 7-10 are scanning electron micrographs of ash rate-carbonized to $873^{O}K$. Figure 7 is a cross-sectional view of an early wood vessel. Tyloses are clearly evident in the vessel, along with fibers and ray parenchyma cells. In Fig. 8 there is evidence of pitting in a late wood vessel and longitudinal parenchyma cells are visible. In addition, ray parenchyma and fibers are also evident. Figure 9 is a radial view of a vessel clearly showing pitting and tyloses and a ray cross field. Almost all of the pit membrane has disappeared at this heat treatment. Figure 10 is a tangential view of rays. Intercellular spaces and end wall pitting in the ray parenchyma cells are clearly visible.

REFERENCES

(1)   A.E. Woodward and J.A. Sauer, "Mechanical Relaxation Phenomena, Physics and Chemistry of the Organic Solid State. Vol. II," 637 (1965).

(2)   G.A. Bernier and D.E. Kline, "Dynamic Mechanical Behavior of Birch Compared with Methyl Methacrylate Impregnated Birch from 90° to $475^{O}K$," For. Prod. Jour., Vol. 18, No. 4, 79 (1968).

(3)   W.L. James, "Effect of Temperature and Moisture Content on: Internal Friction and Speed of Sound in Douglas Fir," For. Prod. Jour., Vol. 11, No. 9, 383 (1961).

(4)   R.F.S. Hearmon, "Theory of the Vibration Testing of Wood," For. Prod. Jour., Vol. 16, No. 8, 29 (1966).

(5)   R.F.S. Hearmon, "The Influence of Shear and Rotatory Inertia on the Free Flexural Vibration of Wooden Beams," British Journal of Applied Physics, Vol. 9, 381 (1958).

(6)   R.E. Pentoney, "Effect of Moisture Content and Grain Angle on the Internal Friction of Wood," Composite Wood, Vol. 2, 131 (1955).

(7)   E. Fukada, "The Vibrational Properties of Wood I.," J. Phys. Soc. of Japan 5, 321 (1950).

(8)   E. Fukada, "The Vibrational Properties of Wood II.," J. Phys. Soc. of Japan 6, 417 (1951).

(9)   B.A. Jayne, "A Non-destructive Test of Glue Bond Quality," For. Prod. Jour., Vol. 5, No. 5, 294 (1955).

(10)   B.A. Jayne, "Indices of Quality ... Vibrational Properties of Wood," For. Prod. Jour., Vol. 9, No. 11, 413 (1959).

(11)   W.J. John and M.M. Lal, "Dynamic Elastic Modulus and Damping Coefficient of Some Indian Timers," Indian Jour. of Phys. 38, 401 (1964).

(12)   W.O. Pillar, "Dynamic Method of ... Determining Curing Properties of an Adhesive in Contact with Wood," For. Prod. Jour., Vol. 16, No. 6, 29 (1966).

(13)   D.G. Miller and J. Benicak, "Relation of Creep to the Vibrational Properties of Wood," For. Prod. Jour., Vol. 17, No. 12, 36 (1967).

(14)   R.E. Pentoney and R.W. Davidson, "Rheology and the Study of Wood," For. Prod. Jour., Vol. 12, No. 5, 243 (1962).

(15)   A.A. Moslemi, "A Study of Moisture Content Gradients in Wood by Vibrational Techniques," Wood Science, Vol. 1, No. 2, 77 (1968).

(16)   D.E. Kline and P.R. Blankenhorn (unpublished data).

(17)   P.R. Blankenhorn, D.E. Kline, and F.C. Beall (unpublished data).

(18) D.E. Kline, "A Recording Apparatus for Measuring the Dynamic
      Mechanical Properties of Polymers," J. of Polymer Science,
      Vol. 22, 449 (1956).

(19) R.E. Taylor and D.E. Kline, "Internal Friction and Elastic
      Modulus Behavior of Vitreous Carbon from 4°K to 570°K,"
      Carbon, Vol. 5, 607-612 (1967).

(20) A.J. Stamm, Wood and Cellulose Science, Ronald Press Company,
      New York (1964).

(21) V.I. Stepanov, B.S. Chudinov, and L.V. Kashkina, Zh. Kim
      ABIPC 39 (10), 8783 (1968).

(22) T. Lim, V. Frosini, V. Zaleckas, D. Morrow and J.A. Sauer,
      "Mechanical Relaxation Phenomena in Polyimide and Poly (2,6-
      dimethyl-p-phenylene oxide) from 100°K to 700°K," (in prepara-
      tion).

(23) C. Schuerch, "Wood Plastization," For. Prod. Jour. 14-9, 377
      (1964).

(24) F.C. Beall and H.W. Eickner, "Thermal Degradation of Wood
      Components," USDA Forest Service Research Paper FPL 130
      (1970).

(25) G.D.M. MacKay, "Mechanism of Thermal Degradation of Cellulose:
      A Review of the Literature," Canada Department of Forestry and
      Rural Development, Forestry Branch, Department Publication
      No. 1201, ODC 813.4 (1967).

(26) P.R. Blankenhorn, G.M. Jenkins and D.E. Kline, "Dynamic
      Mechanical Properties and Microstructure of Some Rate-
      Carbonized Hardwoods," (submitted).

(27) G.A. Bernier and D.E. Kline, "Dynamic Mechanical Behavior of
      a Polyimide," J. of Applied Polymer Science, 12, 593 (1968).

(28) G. Allen, J. McAiush and G.M. Jaffe, Polymer 12, 85 (1971).

(29) J.A. Sauer, Polymer Sci. Symposium 32, 69 (1971).

THE MECHANICAL EXCITATION OF TIE MOLECULES THROUGH CRYSTALLITES

H.H. Kausch

Battelle Institut e.V.   60 Frankfurt (Main) 90

PO Box 900160

ABSTRACT

The problem of the mechanical excitation of tie molecules through crystallites, first treated by Chevychelov, is re-examined in the light of recent ESR investigations of chain breakage. The elastic displacement of stressed chains within crystallites is calculated, including also nonuniform crystal potentials. The maximum axial tension which can be exerted by a defect-free crystallite on a chain is derived as $1.4 \cdot 10^{11}$ dyne/cm^2 (polyethylene, PE) and $2.8 \cdot 10^{11}$ dyne/cm^2 (6-polyamide, 6-PA). Only the latter tension is sufficient to break a chain. After discussing the molecular motion associated with thermal excitation under mechanical stress it is postulated that localized thermal bond excitation outside a crystallite does not interact with elastic chain displacement within crystallites. The contribution by thermal (16 kcal/mole) and mechanical energy (29 kcal/mole) towards activation of scission of a 6-polyamide chain is reported from ESR investigations.

INTRODUCTION

The investigation of the stress-strain temperature relations in high polymers has been both, scientifically rewarding and technologically necessary and fruitful. Together with research into the structure of phases this provides a basis for understanding and predicting the mechanical behaviour of high polymers from their molecular structure (see (1, 2) for references).

A minor problem in this area is that of the mechanical excitation of a macromolecule in a semi-crystalline polymer. The theory of rupture of such a molecule was first treated by Chevychelov (3) who investigated the mechanics of a stressed chain passing from one crystalline region to another through an amorphous one. In particular he studied the elastic interaction between the stressed chain and its rigid crystalline environment which was represented by a sinusoidal potential. The activation energy of chain dissociation then was calculated for a system consisting of the two portions of the chain supported elastically by two crystallites and of those extended chain segments running freely and without interaction through the intermediate amorphous region. Infinite (3) and finite (4) crystal boundaries have been considered by the author.

In the present paper the excitation of chain segments is discussed in the light of recent electron spin resonance (ESR) experiments. The role of molecular motion on the activation process is studied and the Chevychelov problem reformulated accordingly. In addition, consideration is given to the fact that the crystal potential is a function of the chain axis coordinate. The elastic displacements of a stressed chain within such a potential are calculated. They provide a measure of estimating the maximum force which can be transmitted by a stressed chain onto a crystallite before the chain is pulled out of the crystallite.

MODEL

The model used by Chevychelov is a simplified model of a semicrystalline fiber showing crystalline and amorphous regions separated by straight boundaries (Fig. 1). Crossing these boundaries are a few isolated "tie-molecules" (t) which connect adjacent crystallites (c). The crystallites are assumed to be absolutely rigid. The amorphous regions (a) are viscoelastic bodies - of length L_0 - with a compliance J_a considerably greater than that of an extended tie molecule. Any interaction between the tie molecules and the

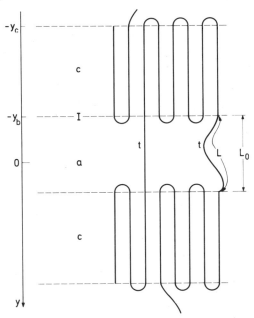

Fig. 1:   Simplified model of crystallites (c) con-
          nected by tie molecules (t)

amorphous region around them is neglected by Chevychelov.

     Application of a macroscopic axial stress σ to
such a system leads to a displacement of the crystallites with respect to each other, which is mostly
determined by the elastic compliance of the amorphous
regions, i.e. $L_0 = L_0(\sigma, J_a)$. As the tie chains are
firmly attached to the crystallites they have to
stretch by about the same amount the amorphous region
does. In view of the very small compliance of molecular
chains this leads to a large stress concentration and
overstressing of extended tie chains.

     Under the action of such stresses the tie chains
try to pull out from the crystal. At the same time the
angles and lengths of main-chain bonds are deformed
within those segments of the tie molecules (t) which
are running through the amorphous regions. Chevychelov
then considered the elastically stretched chain as
fluctuating between states which have small bond lengths
(small excitation of bonds) and a large displacement

of the "crystalline" chain segments with respect to
the crystal potential (large potential energy) and
states with just the opposite properties. He calculated the activation energy of chain scission under
these circumstances (3).

ELASTIC DISPLACEMENT OF CHAINS
WITHIN CRYSTAL POTENTIALS

Theory for Uniform Potentials

Any particular point along a tie-chain may be
characterized by its distance y from the prospective
point of chain scission, where y refers to the unstressed chain. For the molecular point y the displacement under stress is denoted by u = u(y) and
the crystal potential by V(u). The axial stress s
acting upon the chain decreases with increasing |y|
owing to the forces experienced by the chain which is
displaced with respect to the crystal potential V(u).
Equilibrium of forces is established for an infinitesimal section of the chain if:

$$\frac{ds}{dy} = \frac{dV(u)}{du} \tag{1}$$

It should be noted that in deriving this equation no assumptions have been made with respect to the
form of V(u) or the response s(u) of the chain. If one
does, however, assume that the tie-chain segments
within the crystallites resemble an elastic spring,
i.e. if axial stress s is equated to local chain deformation:

$$s = \varkappa \, du/dy \tag{2}$$

we find

$$\varkappa \frac{d^2u}{dy^2} = \frac{dV(u)}{du} \tag{3}$$

where $\varkappa$ is the constant of chain elasticity (not normalized with respect to chain cross-section). The elastic
energy introduced into the system by pulling a chain
a distance of u_o out of the crystal is given by:

$$W_o = \int_0^{u_o} s(u)\ du \tag{4}$$

In an elegant approach Chevychelov (3) replaced the integration over u by that over y:

$$W_o = \int_{-\infty}^{-y_b} s(y)\ \frac{du}{dy}\ dy = \int_{-\infty}^{-y_b} \varkappa\left(\frac{du}{dy}\right)^2 dy \tag{5}$$

The total elastic energy W_o is the sum of two terms: of the energy W_c stored within the chain and of the energy W_p necessary to move the chain against the surrounding potential. As W_c obviously is 1/2 of the term on the right-hand side of equ. (5) it follows that $W_p = W_o - W_c$ has the same value:

$$W_p = \int_{-\infty}^{-y_b} V(u(y))\ dy = W_o/2 = \frac{1}{2}\int_{-\infty}^{y_b} \varkappa\left(\frac{du}{dy}\right)^2 dy \tag{6}$$

and further

$$\frac{1}{2}\ \varkappa\left(\frac{du}{dy}\right)^2 = V(y) \tag{7}$$

From equ. (7) we can derive equ. (3) by differentiation.

For a semi-infinite crystal and a sinusoidal potential $V(u) = v(1-\cos K_o u(y))$ Chevychelov integrated equ. (7) to obtain

$$\frac{1}{2}\left(\frac{du}{dy}\right)^2 = A - \frac{v}{\varkappa}\cos K_o u \tag{8}$$

and, after having determined $A = v/\varkappa$ from $du/dy = 0$ for zero displacement,

$$\frac{du}{dy} = 2\sqrt{v/\varkappa}\ \sin K_o u/2 \tag{9}$$

and

$$u(y) = \frac{4}{K_o}\ \arctan\left\{C\ \exp\left[(K_o\sqrt{v/\varkappa})(y+y_b)\right]\right\} \tag{10}$$

where A and C are integration constants, $-y_b$ the y-coordinate of the crystal boundary, and K_o is 2π

divided by the lattice constant d.

For a finite crystal the boundary condition at
the lower stress end of the chain $(-y_c)$ may become
important. If the chain happens to terminate at $-y_c$ or
continues into the adjacent amorphous region it is
free to respond to any axial forces. The state of
equilibrium, therefore, will only be reached if
$s(-y_c) = 0$, whereas $u(-y_c)$ will, in general, be different from zero. In order to solve this problem we
start from equ. (7). In this case, however, the integration constant A in equ. (8) is smaller than $v/\varkappa$

If a stressed tie-chain folds back at $-y_c$, any
axial displacement there could only occur via deformation of the fold. Such a deformation would cause considerable distortion of the crystal lattice and require
high axial forces. We may, therefore, consider the
boundary condition to be $s(-y_c) > 0$, $u(-y_c) = 0$. In
this case the integration constant A would have to be
larger than $v/\varkappa$. In both cases integration of equ. (8)
leads to elliptic integrals and their inverse functions (4, 5).

Theory for Nonuniform Potentials

It is obvious that in any crystal the potential
shows a certain periodicity. Chevychelov has introduced this periodicity into the form of the potential
(equ. 8), but he has not applied it to the distribution of interacting masses. In this context we will
not discuss the error involved in representing a string
of discrete but identical centers of interaction by a
continuous distribution of forces as done in ref. (5).
In this section, however, we would like to take
account of the fact that in a crystal containing
hydrogen bonds - like that of 6-polyamide - the intermolecular interaction is nonuniformly distributed
along the chain axis (see Fig. 2 for geometry and
notation). It is much stronger for the C=O and N-H
chain groups than it is for the CH_2 groups. The problem
of calculating the elastic displacement of a continuous
chain passing through alternating regions of more or
less intense attraction can be solved by piecewise
application of the inverse elliptic integral functions
using appropriate boundary conditions (5). As a
boundary condition we require u and $s = \varkappa \, du/dy$ to be
steady at all internal boundaries. This is a much more

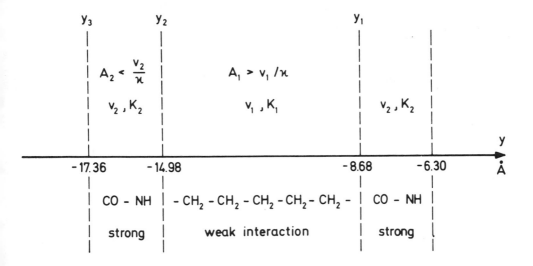

Fig. 2:   Geometry and notations for calculating displacements in 6-polyamide (PA)

stringent condition than that required for chains with
discrete atoms[*]. For the exact derivation of $u(y)$
and $s(y)$ two boundary conditions have to be known. If,
however, s and u tend to zero at the lower stress end
of a tie-chain a relation between s and u exists which
may serve as one boundary condition (5).

For the specific case of a 6-PA chain the following
relation is valid:

$$s(y_3) = 0.425 \; K_2 \sqrt{\varkappa v_2'} \; u(y_3) \qquad (11)$$

$$= 3.31 \cdot 10^{-4} \; u(y_3) \; \text{dyne} / \overset{\text{o}}{\text{A}}$$

(see Appendix for the numerical figures used).

The deformation of the entire system of a stressed
tie molecule within a semi-infinite crystallite,

---

[*]   And it is only under this condition that equ. (6)
is valid. For discrete atoms W_p is generally
different from W_c.

therefore, depends on only one free variable which
may either be tension or displacement at the crystal
boundary $(-y_b)$.

Displacement of Polyethylene (PE) Chains in a PE Lattice

To obtain the axial displacement $u(y)$ for
stressed PE chains equ. (10) was evaluated using the
numerical values given in the Appendix. The resulting
curve $u(y)$ is depicted in Fig. 3 and the corresponding
curve for the chain tension $s(y) = \varkappa\,du/dy$ in Fig. 4.
Some comments are in place here. Displacement and tension are monotonous functions of y. In static equilibrium maximum displacement and maximum tensile stress
of a tie molecule will be found at the crystal surface
$(y = -y_b)$. The largest force (s_0) which can possibly
be applied to a chain is that which causes a displacement of $u = d/2$. This situation is shown in Figs. 3
and 4. If the force becomes any larger than s_0 equilibrium cannot be established, the chain is steadily

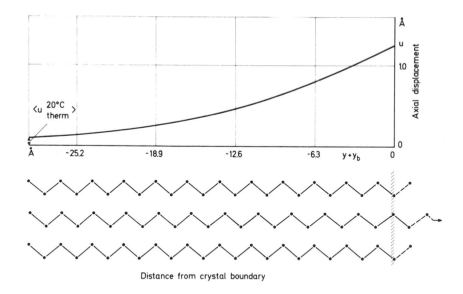

Fig. 3:   Axial displacement of stressed polyethylene
          (PE) chain in PE crystal (u_{therm}: average
          amplitude of thermal vibration)

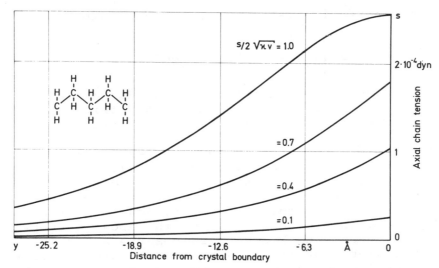

Fig. 4: Axial tension of polyethylene (PE) chain in
         PE crystal for different stresses
         ($s_0 = 2\sqrt{\varkappa v'}$: maximum stress attainable in
         static loading)

pulled out of the crystal (until a change of the
boundary conditions will change the situation). It is
further worth noting that maximum elastic excitation
of a chain at the crystal boundary leads to a distortion of a crystalline layer only 30 Å in thickness.
Beyond that the chain displacement is smaller than the
average amplitude of thermal vibrations at 20°C.
Therefore, the capability of PE crystallites of
holding and exciting tie molecules does not depend on
their thickness (if it exceeds 30 Å).

The maximum chain tension in static equilibrium
(equ. 2 and 9) is:

$$s_o = \varkappa (du/dy)_{u=d/2} = 2\sqrt{v\varkappa'} = 25.8 \cdot 10^{-5} \text{ dyne} \qquad (12)$$

This force s_0 which pulls an infinitely long chain is
only 8.2 times larger than the force $K_0 vd$ (= $3.14 \cdot 10^{-5}$
dyne) which is necessary to pull a single monomer unit
out of the crystal.

Dividing s_{oo} by the average cross section of a molecule
(A = 18.24 Å^2) we arrive at the maximum axial stress

Υ_o which a perfect crystallite is able to exert on a tie molecule:

$$\Upsilon_o = 1.41 \cdot 10^{11} \; dyne/cm^2 \tag{13}$$

This value should be compared with the "strength" of a PE chain for which we may assume about twice the strength of a 6-PA chain, i.e. $4 \cdot 10^{11}$ dyne/cm^2 (6).

Comparison shows that under slow excitation a PE chain tends to be pulled out of a crystallite rather than breaking. At this point we will not discuss the effect of dynamic loading, lattice defects, or chain length. It should be noted, however, that some recent ESR investigations seem to support the above assumption, since Becht and Fischer did observe a barely detectable amount of free radicals in loaded PE fibers (indicating little chain scission) and a rather severe break-up of crystallites (7).

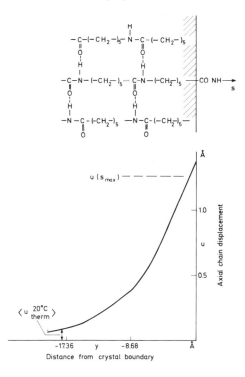

Fig. 5: Axial displacement of stressed 6-polyamide (PA) chain in, PA lattice. The displacement $u(s_{max})$ belonging to the maximum stress is shown.

Displacement of Polyamide (PA) Chains in a PA Lattice

The displacement u calculated for 6-polyamide chains according to the inverse elliptic integral functions proposed by Kausch and Langbein (5) and using the data shown in the Appendix is illustrated in Fig. 5. Again it is noted that $u(y)$ drops off readily with y and reaches the level of thermal vibrational amplitudes at about 20 Å from the crystal boundary. The inhomogeneity of the crystal potential becomes drastically apparent from Fig. 6. The strong interaction of the CONH groups with each other through hydrogen bonds leads to a rapid decrease of axial chain tension in exactly these parts of the stressed tie-chain. The curve shown in Fig. 6 represents a

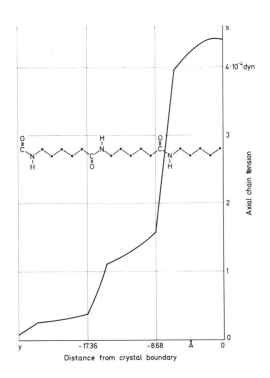

Fig. 6:   Axial tension of polyamide (PA) chain in PA lattice. Position of carbonamide groups of PA lattice is shown to scale in the insert. For the calculations the crystal boundaries have been assumed to be sharp

maximum chain tension of $4.4 \cdot 10^{-4}$ dyne; this is about
1.7 times the greatest possible tension exerted on a
PE chain. It should be pointed out that the value of
the maximum of the chain tension depends the position
of the boundary and varies between 4 and $5 \cdot 10^{-4}$ dyne
(5). A force of $K_2 v_2 d = 2.6 \cdot 10^{-4}$ dyne is necessary for
pulling out a single CONH group. This means that the
major part (65 to 52 %) of the force to pull out a
whole PA chain is expended in breaking the hydrogen
bonds of a single CONH group. If s_{max} is divided by
the chain cross section (17.6 $\overset{o}{A}^2$) a stress of
$2.84 \cdot 10^{11}$ dyne/cm^2 is obtained. The strength of a PA
chain, however, is only $2.2 \cdot 10^{11}$ dyne/cm^2 (6). This
explains the great probability for chain scission in
6-polyamide and 66-polyamide (7). It should be noted
that the critical strength of a capron chain as calculated by Chevychelov on the basis of different data
(4) was given as $5.4 \cdot 10^{-2}$ "atomic units" or
$2.54 \cdot 10^{9}$ dyne/cm^2 - in disagreement with general
experience (1, 6, 7).

MOLECULAR MOTION AND CHAIN TENSION

        After having thoroughly investigated the clamp-
like function of crystallites with respect to tie
molecules we have to discuss the mechanical excitation
of tie molecules, their vibration, relaxation and
scission. The mechanical excitation has become clear
by the foregoing discussion: the displacement $G J_a L_o$
of the crystal boundaries leads to a stretching of the
tie molecules and to their being pulled out from the
crystal by an amount of $2u(s)$. In static equilibrium
we have

$$L_o + G J_a L_o = L + Ls/\varkappa + 2u(s) \qquad (14)$$

        We may say that the system reacts with a compliance
which is that of the elastic chain $(L/\varkappa)$ increased by
the compliance $2u/s$ of the "clamp". The clamp compliance $2u/s$ is a function of the chain displacement
varying within very narrow limits. For the system PE
chain/crystal these limits are $6.2 \cdot 10^{3}$ $\overset{o}{A}$/dyne (small
displacements) and $1 \cdot 10^{4}$ $\overset{o}{A}$/dyne (maximum elastic displacement of $d/2$). For the system PA chain/crystal the
equivalent figures are $3 \cdot 10^{3}$ $\overset{o}{A}$/dyne and $5.5 \cdot 10^{3}$ $\overset{o}{A}$/dyne,
respectively.

The term $L/\varkappa$ for a 50 $\overset{\text{o}}{A}$ segment of a PE or PA chain has a value of $1.47 \cdot 10^4$ $\overset{\text{o}}{A}$/dyne. The compliances of the chains are larger than the respective clamp compliances, but only by a factor of between 1.5 and 5. Clamping, therefore, does not appear to be very rigid.

The stressed tie-chain is assumed to be in thermal contact with its surrounding, which means that the chain atoms take part in the exchange of thermal energy through quantized units of vibrational energy, the phonons. The vibrational excitation of a particular group may be characterized by the number and type of phonons simultaneous exciting that group. It is worthwhile discussing the type of molecular motion described by different phonons. In particular C-C-bond stretching (wave number ~ 1000 cm^{-1}, period of oscillation $1/\nu_s = 0.03 \cdot 10^{-12}$ sec) and segment bending, torsion and translation ($\bar{\nu}$ less than 200 cm^{-1}) will be considered.

The dissociation of a bond requires a localized excitation, an out-of-phase motion of the two bound atoms of an amplitude high enough that the energy barrier can be overcome. Obviously phonons with large k values contribute more towards an out-of-phase motion than phonons with, say, $k < \bar{\pi}/4d$. The latter describe an in-phase motion of neighboring chain atoms. If longitudinally polarized such a vibration effects the stress-induced elastic displacement of the chain with respect to the crystal potential. It should be observed, however, that this in-phase vibration of a group of chain atoms does not change any static tensions acting between those atoms. The dissociation, therefore, must be caused by statistical coincidence of a large number of phonons with k close to $2\bar{\pi}/d$. The velocity of these phonons is small and their propagation along an "amorphous" chain disturbed by phonon-phonon inter-action and elastic and/or inelastic scattering at defect sites (8). Thus, their relaxation length is less than 10 $\overset{\text{o}}{A}$. We postulate, therefore, that in sta-tic loading of a chain the thermal bond excitation does not interact with the elastic excitation of the chain terminals.

We are now able to discuss the thermo-mechanical dissociation; a bond and, for that matter, a chain scission if the thermal bond energy is equal to or larger than the dissociation barrier. In a planar

zigzag chain of tetrahedral bond angles the height U
of the dissociation barrier is comparatively little
affected by the amplitude and phase of the stretching
vibrations of the neighboring (C-C)bonds; U decreases,
however, with chain tension s. Also, any skeleton
vibration of the chain which leads to a bond angle
deformation will increase or decrease the chain tension and thus the dissociation barrier U. In as much
as such changes of the dissociation energy average
out within a period of, say, 10^{-10} sec they will
usually be considered together with bond excitation.
We are thus left with only three quantities affecting
the probability of bond dissociation: thermal excitation, axial chain tension, and conformational changes
influencing either s or U, or both. The latter changes
also include relaxation processes.

From the theory of rate processes (9) the most
commonly used equation for the probability of bond
dissociation $1/\tau$ is derived as

$$1/\tau = \omega_b \exp\left\{-(U_o - \beta\dagger)/RT\right\} \qquad (15)$$

where ω_b is a frequency representative of the fluctua-
tion of thermal excitation (i.e. not necessarily
identical with ν_s if skeleton vibrations of a frequency smaller than ν_s are also considered to contribute to
thermal excitation). The activation energy for bond
scission, U_o, refers to an unstressed chain, $\dagger$ is
chain tension s per cross section of the chain, and R
and T are gas constant and absolute temperature. The
quantity ß describes the effect of uniaxial tension
on the height of the potential barrier U, ß has the
dimension volume/mole.

Numerical data were obtained by electron spin
resonance (ESR) investigation of thermo-mechanical
dissociation of 6-polyamide chains (6, 7, 10, 11, 12).

Thus, ß was determined to be $1.32 \cdot 10^{-10}$
kcal mole^{-1} dyne^{-1} cm^2. This value is equivalent to
an activation volume of 9.2 A^3 per broken bond. The
"strength" of the weakest bond of the chain - an
(NH)-C-C- or (CO)-C-C-bond - was found to be
$22 \cdot 10^{10}$ dyne/cm^2. Hence, the stress-dependent reduction of the activation energy U is $\beta\dagger = 29$ kcal/mole.
As U_o is 45 kcal/mole an amount of 16 kcal/mole has to
be contributed by the thermal excitation towards
breaking a bond. It is interesting to note that the

average thermal excitation of a CH_2 group is only
0.83 kcal/mole (12).

        As a final point we would like to consider the
effect of relaxation processes on the rate of chain
scission. It follows from the model employed that
three different processes will be of importance:

1.   The change of conformation of the amorphous
      segments of stressed tie molecules involving
      axial extension or contraction

2.   Any processes leading to the pull-out of
      stressed chains from crystallites

3.   The rotation or translation of highly stressed
      crystallites.

        Below T_g the first process occurs in 6-polyamide
and 66-polyamide at a very small but noticeable rate,
which determines the dependence of the number of bond
breakages from the rate of loading. This can be stated
with some certainty only for those two polymers as ESR
experiments have shown that neither translation of
individual highly stressed crystallites nor a break-up
of crystallites did occur (6, 7, 10).

        The second process, the release of stressed tie
molecules from crystallites, plays a major role in
static loading of PE, polypropylene (PP), 12-PA and
polyethylene terephthalate (PET). This was concluded
by Becht and Fischer from the almost unobservable
number of free radicals formed during mechanical excitation, from the absence of molecular-weight changes,
and from the decrease of the heat of fusion of strained
samples of 12-PA, PP and PET.

        The third process is of importance in the loading
of unoriented spherulitic materials. It is particularly
noticeable in the major rearrangements involved in
yielding.

        In all three cases the conformational or struc-
tural changes tend to release the axial stress applied to
highly loaded chain segments. Chain scission, then,
has to compete with stress relaxation. The calculations
in section 3 have shown that there is only a narrow
margin left for the occurence of chain scission in
static loading. In rapid loading this margin may be

considerably widened and leads to a measurable number
of chain breakages in other semicrystalline (PE, 12-PA)
or even amorphous polymers (polycarbonate) as well.

LITERATURE

(1)   Sauer, J.A., A.E. Woodward, Chapt. 3 in Polymer
       Thermal Analysis, II., P.E. Slate, Jr. and
       L.T. Jenkins (Eds.), Dekker, Inc. (New York 1970)

(2)   Sauer, J.A., J. Polym. Sci. C, $\underline{32}$, 69 (1971)

(3)   Chevychelov, A.D., Polymer Science USSR $\underline{8}$, 49
       (1966)

(4)   Chevychelov, A.D., Polymer Mechanics (Mekhanika
       Polimerov), $\underline{2}$, 415 (1966)

(5)   Kausch, H.H., D. Langbein, to be published
       phys. stat. solidi (a) (1972)

(6)   Kausch, H.H., J. Becht, Rheologica Acta, $\underline{9}$, 137
       (1970)

(7)   Becht, J., H. Fischer, Kolloid-Z. u. Z. Polymere
       $\underline{229}$, 167 (1969) and ibid. $\underline{240}$, 766 (1970)

(8)   Baur, H., Kolloid-Z. u. Z. Polymere, $\underline{247}$, 753
       (1971)

(9)   Tobolsky, A., H. Eyring, J. Chem. Phys., $\underline{11}$, 125
       (1943)

(10)   Becht, J., H. Fischer, Angew. Markom. Chemie,
       $\underline{18}$, 81 (1971)

(11)   Kausch, H.H., Kolloid. Z. u. Z. Polymere, $\underline{247}$,
       768 (1971)

(12)   Baur, H., B. Wunderlich, Fortschritte der Hoch-
       polymeren Forschung $\underline{7}$, 388 (1970)

APPENDIX

Data used for calculations

Polyethylene:

Distance between alternating
atoms $\quad d_1 = 2.52$ Å

$$K_1 = 2\pi/d_1 = 2.5 \text{ Å}^{-1}$$

Chain elasticity constant $\quad \mathcal{x} = 34\cdot10^{-4}$ dyne

Crystal potential depth $\quad v_1 = 5\cdot10^{-6}$ erg/cm
  (derived from a
    solubility parameter: $\quad \delta^2 = 64$ cal/cm^3)

6-Polyamide:

$d_1 = 2.52$ Å

$d_2 = 2.38$ Å; $\quad K_2 = 2\pi/d_2 = 2.64$ Å^{-1}

$\mathcal{x} = 34\cdot10^{-4}$ dyne

$v_1 = 5\cdot10^{-6}$ erg/cm

$v_2 = 40.7\cdot10^{-6}$ erg/cm (derived from a solubility
parameter: $\delta^2 = 184$ cal/cm^3)

ANOMALOUS MECHANICAL BEHAVIOUR IN LINEAR POLYETHYLENE BELOW 20°C

J. W. Cooper and N. G. McCrum

Department of Engineering Science

Oxford University, Oxford, England

ABSTRACT

The temperature dependence of logarithmic decrement Λ and shear modulus G' is measured at 0.67 Hz at temperatures from 20° down to -196°C for a specimen of linear polyethylene of density 0.936 gm/cc (Hi-fax 1900). The observed relaxations depend on the thermal history. If the specimen is quenched from room temperature to -196°C and measurements taken continuously as the temperature rises, then the relaxations observed are γ_{II} (-160°C), γ_I (-120°C), and β(-20°C). If the specimen is cooled slowly from one temperature to another and held at constant temperature for times of the order of one hour prior to the measurement of Λ and G', then only the dominant γ_I relaxation is resolved, although a shoulder is observed on the low temperature side of the γ_I peak in the region of the γ_{II} peak of the quenched specimen. It is concluded that quenching from room temperature into liquid nitrogen magnifies in linear polyethylene relaxations which are barely detected or undetected in specimens which have been slowly cooled. The acceptance of the β relaxation as a secondary relaxation of linear polyethylene is of considerable conceptual significance since it eliminates a long standing paradox presented by copolymerization studies.

INTRODUCTION

The weight of experimental evidence supports the assignment of the low temperature γ relaxation in linear polyethylene (LPE) to the amorphous fraction. For instance, a correlation is observed between the γ peak[1] (in logarithmic decrement Λ) at -100°C (10 Hz) and an abrupt change in thermal expansion coefficient in the region of -130°C[2-4]. This correlation is particularly forceful because the magnitude of the γ relaxation[5] and the change of thermal expansion coefficient[4] are observed to increase with increasing amorphous fraction. These and other experiments on bulk linear polyethylene as well as experiments on crystal mats[6] have yielded renewed support to the hypothesis of Willbourn[7] that the γ relaxation is the mechanical relaxation associated with the glass transition (T_g) of the amorphous fraction.

There are, however, facts which are not resolved by the work quoted in the preceeding paragraph. *First* , what is the origin of the multiple damping peaks observed sometimes below the main γ peak? Occasionally a definite structure is resolved, but more often a broad shoulder appears on the low temperature side of the γ peak. *Second*, according to several authors the γ relaxation is due to stress induced reorientation of defects within the lamellae[8,9]. The evidence for this detailed crystalline mechanism is based on experiments on polycrystalline n-alkanes[10] and single crystal mats[8]. Can this view be reconciled with the overwhelming evidence for the amorphous mechanism of the γ relaxation? For instance, is it possible that both types of relaxation occur, the amorphous component being normally of greater magnitude except in polycrystalline n-alkanes and unannealed single crystal mats? *Third*, the acceptance of the γ transition* as the T_g of LPE leads to the following paradox[6]. This is that it must follow by the same reasoning, based on experimental evidence from mechanical damping (including variations in branch content[11], copolymer content[12], chlorine content[13]) and x-ray scattering[6,14], that the β transition is the T_g of branched polyethylene (BPE). Yet in addition to the β relaxation, BPE exibits a γ relaxation which is clearly an exact analogue of the γ relaxation in LPE[11,13]. It may be asked how the addition of one or two small side branches per 100 carbon atoms can (1) cause the γ transition to cease to be the glass transition; (2) cause the β transition to become the glass transition. This type of dependence of T_g on chemical composition (see for instance Illers[15]) is quite unprecedented. The acceptance of the γ transition as the T_g of LPE and the β transition as the T_g of BPE implies an absurd dependence of T_g on comonomer content which cannot be reconciled on the basis of the classical concept of the glass transition.

---

* Following normal usage, we restrict the term 'transition' to the volumetric transition and the term 'relaxation' to the associated frequency dependent mechanical relaxation.

In the experiments described below we have attempted to remove these inconsistencies in the experimental evidence by studying in LPE the effect of quenching. It is known that quenching from well above to well below T_g in amorphous polymers produces marked and well understood effects on the mechanical properties[16,17,18]. In our experiment the initial programme (later extended) was to quench from room temperature (well above the β transition) to liquid nitrogen temperature (well below the γ transition) and to observe the effects on Λ and the shear modulus G'. It was anticipated that if T_g is at -130°C marked annealing effects would be observed in that region of temperature. Control experiments were also performed in which the specimen was cooled slowly from room temperature and maintained at the measuring temperature for periods of at least one hour before measuring Λ and the shear modulus G'.

<u>EXPERIMENTAL</u>

A specimen of linear polyethylene of size 8.5 x 1.0 x 0.10 cm. was machined from a sheet of Hi-fax 1900 obtained from the Hercules Powder Company. According to manufacturer's data this polymer is of extremely high molecular weight (weight average in the range 2 to 5 million). The density of the specimen was 0.936 g/cc at 23°C. Measurements of Λ and G' were obtained using an inverted torsion pendulum operating at a constant frequency of 0.67 Hz. The specimen was surrounded by a copper jacket which was thermally insulated by a sheath of polyurethane foam. A coil of copper tubing was soldered to the outer surface of the copper jacket. The temperature of the specimen was controlled by regulating the flow of cold nitrogen gas through the coil using electronic control equipment. The temperature was measured by three copper-constantan thermocouples placed about 1 cm. from the specimen and arranged so as to monitor the magnitude of the temperature gradient along the specimen. Values of Λ and G' were computed using standard equations[19]. G' was not corrected for small changes in the geometrical form factor due to thermal expansion: the magnitude of the correction is of the order of 5% at liquid nitrogen temperature.

Two different thermal histories were studied. These are identified as non-equilibrium and equilibrium. In a non-equilibrium experiment the specimen (mounted in the torsion pendulum) was quench-cooled* from room temperature by a rapid immersion in liquid nitrogen. This was achieved by placing the insulated copper jacket immediately below the pendulum, filling it with liquid nitrogen and then abruptly

---

* The term 'quench-cooled' is often used for a specimen quenched from the melt. In this paper the term refers to specimens quench-cooled from room temperature to liquid nitrogen temperature by immersion in liquid nitrogen.

raising it (into its normal position) so as to surround the
specimen. The liquid nitrogen boiled off rapidly. The temperature
was then maintained at -180°C by passing cold nitrogen gas through
the cooling coils. After a period of about 10 minutes the flow of
cold nitrogen gas was stopped and the specimen allowed to warm up
due to heat leakage through the thermal shield of the apparatus.
Measurements of Λ and G' were obtained "on the run" as the temperature rose. The initial rate of temperature rise was 1.5°C/minute
dropping to 0.1°C/minute near room temperature. In the equilibrium
experiment the specimen was held at constant temperature for at
least one hour at each temperature before Λ and G' were measured.

RESULTS

     The results of the non-equilibrium and equilibrium experiments
are shown in Figure 1. It is to be noted that the non-equilibrium
experiment yields three apparent relaxations which are labelled,
for purposes of reference,β,γ_I and γ_{II} following Illers[20]. The
equilibrium experiment yields the γ_I relaxation.

     The two experiments shown in Figure 1 differ in two respects.
In the non-equilibrium experiment (1) the specimen was quenched to
-196°C from room temperature (2) the data points were taken "on the
run" as the temperature rose. In the equilibrium experiment (1)
the specimen was not quenched and (2) the data points were taken
after the specimen had been held at each temperature for a time in
excess of one hour. It seemed likely that both factors were
significant. This was supported by a pseudo-equilibrium experiment
in which (1) the specimen was cooled slowly over a period of three
hours from 20°C to -196°C (2) the data points were taken "on the run"
as the temperature rose due to heat leakage through the thermal
shield. The measured values of Λ and G' lay between the curves
shown in Figure 1. Nevertheless, it seemed important to establish
in greater detail the interrelation between quenching and time-at-
temperature and this was performed in the following experiment.

     The specimen was quenched to -196°C, held at that temperature
for ten minutes and then the cooling flow of cold nitrogen gas cut
off (as in a non-equilibrium experiment). Heat leakage then caused
the temperature to rise to -167°C. The thermostat was at once set
to maintain this temperature but in fact over the ensuing period of
eighty minutes the temperature drifted down to -173°C. During the
fairly rapid rise in temperature (-180°C to -167°C) and during the
slow drift down (-167°C to -173°C) measurements of Λ and G' were
obtained and are shown in Figure 2 (cycle I). It will be seen that
the points form a loop. The temperature was then lowered to -180°C.
The specimen was next permitted to warm up (again due to the natural
heat leak through the thermal shield) to -137°C and was annealed at

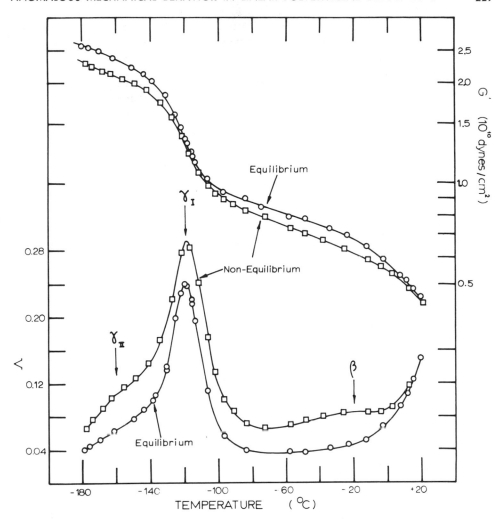

Figure 1. Temperature dependence of logarithmic decrement Λ and shear modulus G' for two thermal histories. (i) <u>Equilibrium</u>: specimen (not quenched) held at each temperature for one hour before measurement. (ii) <u>Non-equilibrium</u>: specimen quenched to liquid nitrogen temperature, measurements then being taken as the specimen warmed up.

that temperature. Measurements of Λ and G' were obtained during the rise in temperature to -137°C and again during the annealing period. The points are shown in Figure 2 (cycle II). It will be seen that during the annealing period Λ fell and G' rose in magnitude. This experiment was continued, seven temperature cycles in all being observed, Figure 2:   in each case points were obtained as the temperature drifted up from $-180^{\circ}C$ to the annealing temperature and at the annealing temperature. Many of the points have been omitted to preserve the clarity of the graph.

The pattern of behaviour is clear. The equilibrium and non-equilibrium experiments of Figure 1 are limiting cases. Consider a specimen which has been annealed at T_1 and is to be subjected to a temperature cycle from $-180^{\circ}C$ to the next annealing temperature

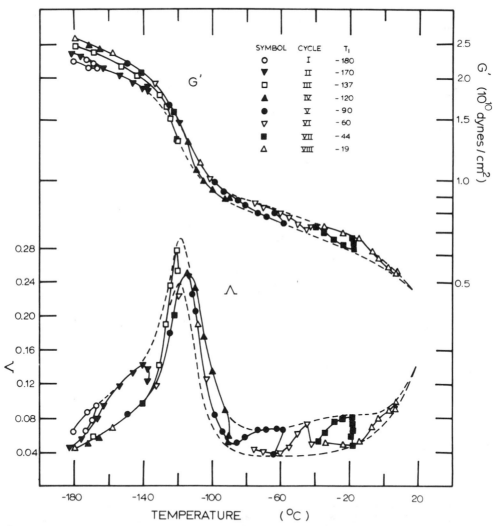

SYMBOL	CYCLE	T_1
O	I	-180
▼	II	-170
□	III	-137
▲	IV	-120
●	V	-90
▽	VI	-60
■	VII	-44
△	VIII	-19

Figure 2: Temperature dependence of logarithmic decrement Λ and shear modulus G'. See text for description of thermal history. The dashed curves, which form envelopes to the data, are the equilibrium and non-equilibrium curves of Figure 1.

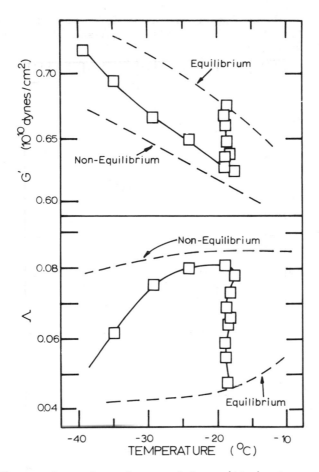

Figure 3: Temperature dependence of logarithmic decrement Λ and
shear modulus G' (enlarged scale plot of cycle VII, Figure 2).
See text for description of thermal history. The arrows indicate
the time sequence.

T_2 (above T_1). As the temperature rises from -180°C towards T_1
the data fall close to the equilibrium curve. In the vicinity
of T_1 the data points cross over from the equilibrium curve to
the non-equilibrium curve. At somewhat above T_1 the data fall
close to the non-equilibrium curve. The loop occurs because
during the anneal at T_2 the data points move isothermally from the
non-equilibrium to the equilibrium curve. This effect is shown in
greater detail for cycle VII (T_1 = -44°, T_2 = -19°C) in Figure 3.
When the specimen temperature exceeds T_1 the values of Λ and G' move
towards the non-equilibrium curves: during the annealing period at
T_2 the values of Λ and G' move back towards the equilibrium curves.
The time dependence of Λ and G' for this particular annealing period

(T_2 = -19°C) is shown in Figure 4. It will be seen that for
annealing times less than 10^4 seconds both Λ and G' vary approximately linearly with log t.

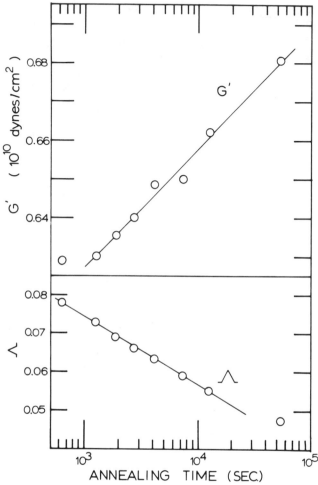

Figure 4: The isothermal dependence of logarithmic decrement
and shear modulus G' on log time at annealing temperature
T_2 = -19°C (cycle VII, Figure 2).

DISCUSSION

(a) Multiple Damping Peaks

In the description of experimental conditions, authors do
not normally record thermal history. This is particularly true
at low temperatures where, until now, it was often thought that
thermal history could not be of importance. It will be seen
from Figure 1 that it will now be necessary for a reappraisal of

much of the evidence in the literature. Many workers perform
experiments by cooling, sometimes rapidly, to liquid nitrogen
temperature and then measuring Λ and G' as the specimen warms
up to room temperature. The recorded spectrum under these circumstances will then depend not only on the properties of the specimen
(crystallinity, orientation, etc.) but also on the original rate of
cooling and on the rate of heating. This is not to say that
experiments performed in this way cannot be of value (see sections
4(b), 4(d)). But when interpreting them it must be understood
that the thermal history of the experiment has a role which must
be accounted for. We mention two important cases in which thermal
history may have caused the resolution of a relaxation which would
be unresolved in an equilibrium experiment. The first is the β
peak observed in a specimen of polymethylene[7]. The second is the
observation of γ_{II} and γ_{III} peaks below the γ_{I} peak[20].

(b) Stress Induced Reorientation of Defects

The evidence in favour of this mechanism is due particularly
to Illers[10], Sinnott[8], Hoffman, Williams and Passaglia[9] and Shen
and Cirlin[21]. Of the somewhat meagre evidence, that shown by
Sinnott (reference 8, Figure 12) is probably the most formidable.
If the temperature of the γ relaxation in single crystal mats and
of the low temperature relaxation in n-alkanes is plotted against
the logarithm of the number of CH_2 groups (n) between the fold
surfaces (polyethylene) or in the molecule (n-alkanes) then the
data for both materials falls on one straight line. As n increases
from 10 to 300 the relaxation temperature increases systematically
from ~ -183°C to -120°C. The obvious conclusion is that it is the
crystal phase which relaxes, a hypothesis which is well argued in
detail both by Sinnott[8] and by Hoffman, Williams and Passaglia[9] .
Shen and Cirlin[21] show that single crystal mats with the amorphous
fraction supposedly entirely eliminated by nitric acid etching
still exhibit a γ maximum.

Of the many arguments against this interpretation we find
those of Stehling and Mandelkern[4] the most persuasive. For
instance, it is possible that the n-alkane relaxation is due to
molecularly dispersed paraffin molecules in the polystyrene matrix.
Furthermore, there is no evidence of a low temperature relaxation
in the single crystal of n-eicosane studied by Crissman and
Passaglia[22]. Nevertheless, no satisfactory rationale of all the
data exists at present.

It is to be noted that Sinnott by quenching from 125°C into
Dry-ice-acetone slurry failed to increase the magnitude of the
γ peak over that of a slowly cooled specimen. This was taken to
be evidence against the defect hypothesis. We do not feel that the

increased size of the γ_I peak (and the existence of the γ_{II} peak)
produced by quenching in our experiments *inevitably* supports the
defect hypothesis. The major argument against such an explanation
is that quenching seems to increase the damping at all temperatures.
It is clear nevertheless that quenching produces a fairly
pronounced γ_{II} peak in addition to raising the general level of
mechanical loss at all temperatures. The quenched-in crystal
defect hypothesis is therefore attractive, yet without further
evidence it remains a stimulating hypothesis with little experimental support.

It may well be asked whether the slight shoulder observed in
the annealed specimen in the region below the γ_I peak (Figure 1)
could be eliminated by additional annealing. This is to say, even
in the so called equilibrium experiment the specimen may not be
truly in equilibrium. This is an important point which requires
further elucidation. It could well be that further annealing at
each measuring temperature would make the damping peak observed
in the equilibrium experiment more symmetrical. It will be of
interest to determine the relationship between the γ_{II} peak
observed in our non-equilibrium experiment in LPE and the subsidiary
low temperature peaks (γ_I, γ_{II}) observed in polyoxymethylene by
Papir and Baer[23]. Evidence for these low temperature relaxations
has also been obtained in dielectric relaxation by Hideshima and
co-workers[24]. Papir and Baer[23] have strong evidence for a
crystalline relaxation at $48°K$ (~ 1 Hz). The magnitude of this
relaxation has found to increase with increasing crystal fraction.

(c) Mechanism of the Quenching Effect

It has been known for some time that quenching effects are
easily observed in the vicinity of the glass transition Tg. This
follows in particular from the work of McLoughlin and Tobolsky[16]
on polymethyl methacrylate and Kovacs, Stratton and Ferry[17] on
polyvinyl acetate. At the phenomenological level the effect is
understood: if an excess free volume Δv_o is frozen in over and
above the equilibrium free volume v_o, and if, for instance,
$\Delta v_o/v_o = 0.003$ then the relaxation time according to the Doolittle
equation will be altered by a factor of about 400[17]. Thus very
small changes in total volume have a substantial effect on the
viscoelastic properties. We examine first the possibility that an
analogous mechanism may explain the observed quenching effect in
LPE.

For one class of amorphous polymers, the polyalkyl methacrylates,
there is abundant evidence to support the proposal of Simha and
Boyer[25] that there is a close connection between the low temperature
mechanical loss processes[26,27] and excess free volume observed below

T_g. According to this view one form of glass (Glass I) forms from the supercooled liquid at Tg, the Glass I transition. At a considerably lower temperature Glass II is formed from Glass I at the Glass II transition[28],[29]. The change in thermal expansion at the Glass II transition is much smaller than at the Glass I transition. Similarly the maximum change in Λ at the Glass II relaxation is much smaller than at the Glass I relaxation. It is apparent that even below Tg the free volume not only depends on temperature but also depends on thermal history and on time[28]. It follows from this premise, considering the large observed change in thermal expansion coefficient at the γ transition[4], that the free volume of the amorphous fraction in LPE will also depend on thermal history and on time. If the molecular motion responsible for the γ relaxation depends on free volume[30] then clearly quenching and annealing will affect Λ and G'.

There are two difficulties with this interpretation (the quenched-in free volume theory). The first (and less severe) concerns the fact that annealing in the experiments shown in Figure 2 takes place at temperatures as low as -167°C, approximately 40°C below the transition temperature, -130°C. We have other evidence which indicates that annealing takes place at liquid nitrogen temperature, approximately 60°C below transition. Is it possible that free volume effects could be observed under these circumstances?

The second difficulty with the quenched-in free volume theory is that if the γ transition is the one and only glass transition then no quenching effects should be observed above -130°C. But as can be seen in Figures 3 and 4 pronounced annealing of a quenched specimen is observed at -19°C. That is to say, a specimen of LPE is just as susceptible to quenching above as below -130°C. This attack is partially deflected if the β transition at -30°C is accepted as a secondary transition in LPE: in our view this is supported in any case by the data shown in Figure 1 (see Section 4d). However, the pronounced effects shown in Figure 3 were obtained at an annealing temperature of -19°C somewhat above the approximate β transition at -30°C. Of clearer significance is a recent observation that annealing effects in LPE analogous to those shown in Figure 3 are observed at temperatures as high as 40°C[31] (and which are not due to recrystallization). That is to say, quench cooled specimens of LPE exhibit annealing effects over the whole range of temperature from -196°C up to +40°C. It follows from this that the quenched-in free volume theory cannot explain *all* the results. In the next section we discuss whether or not in the region of the β relaxation the peak exhibited in the non-equilibrium experiment (Figure 1) may be caused by quenched-in free volume. This possibility is not excluded. Nevertheless, the fact that non-equilibrium experiments always yield higher Λ (and lower G') than equilibrium experiments and that this occurs at all

temperatures implies that there must be some other mechanism
effective at all temperatures. One theory currently under
investigation in our laboratory attributes this additional effect
to the relaxation (during the period of annealing) of thermal
stresses set up by the quench to liquid nitrogen temperature.
This work will be described in detail elsewhere.

(d) The Amorphous Transitions of Linear Polyethylene

It is well known that the β relaxation in BPE decreases in
magnitude as the side branch content decreases[11]. It will be seen
that in the equilibrium experiment, Figure 1, the β peak if present,
is an unresolved shoulder to the α peak. However, in the non-
equilibrium experiment a β peak is clearly resolved. This behaviour
is reminiscent of the β relaxation in polyoxymethylene ($0^{\circ}C$, 1 Hz)[32].
Both relaxations can be enhanced in size by quenching and essentially
eliminated by annealing. The β relaxation in POM is also the
relaxation which is affected by copolymerization: as the $\{CH_2-CH_2-O\}$
content is increased the β relaxation increases in size[32]. This
effect is identical to the increase in the size of the β peak in
polyethylene as chlorination is increased[13] or as the vinyl acetate
content of copolymers is increased[12].

For the foregoing reason it is probably correct to consider
the β relation as a small secondary amorphous relaxation of LPE.
If this is accepted then the paradox outlined in the second paragraph of Section 1 is resolved in the following way. Both LPE and
BPE have more than one amorphous transition. In LPE the dominant
transition is at $-130^{\circ}C$ but the transition at $-30^{\circ}C$ although small,
is not negligible. In BPE the dominant transition is the β transition at $-30^{\circ}C$ with the γ transition of secondary importance. The
inherent characteristics of the γ and β transitions are unaltered
by the addition of a low concentration of side branches, which merely
controls the transition magnitudes. If this is accepted, it is
clear that non-equilibrium experiments can be valuable in resolving
small relaxations which remain unobserved under equilibrium conditions.

CONCLUSIONS

Quenching from $20^{\circ}C$ to $-196^{\circ}C$ enhances in LPE the magnitude of
mechanical loss at all temperatures. Two relaxations, β and γ_{II},
are detected in quenched specimens which are unresolved in LPE which
is cooled slowly from $20^{\circ}C$ to the measuring temperature. The γ_I
relaxation is the dominant relaxation in quenched and in slow cooled
specimens. The acceptance of the β relaxation as a secondary
relaxation of LPE eliminates a long standing paradox presented by
copolymerization studies. The origin of the γ_{II} relaxation is

unknown.   The γ_I relaxation is the major amorphous relaxation of LPE.

REFERENCES

1. K. Wolf and K. Schmieder, La Ricerca Scientifica, (Suppl.A) 25, 732 (1955).
2. M.L. Dannis, J.Appl.Polym.Sci., 1, 121 (1959).
3. P.R. Swan, J.Polym.Sci., 56, 403 (1962).
4. F.C. Stehling and L. Mandelkern, Macromolecules, 3, 242 (1970).
5. R.W. Gray and N.G. McCrum, J.Polym.Sci. A-2, 7, 1329 (1969).
6. E.W. Fischer and F. Kloos, J.Polym.Sci. B, 8, 685 (1970).
7. A.H. Willbourn, Trans.Farad.Soc. 54, 717 (1958).
8. K.M. Sinnott, J.Appl.Phys., 37, 3385 (1966).
9. J.D. Hoffman, G. Williams and E. Passaglia, J.Polym.Sci., C, 14, 173 (1966).
10. K.-H. Illers, Rheol.Acta, 3, 194 (1964).
11. D.E. Kline, J.A. Sauer and A.E. Woodward, J.Polym.Sci., 22, 455 (1956).
12. L.E. Nielson, J.Polym.Sci., 42, 357 (1960).
13. K. Schmieder and K. Wolf, Kolloid-Z. Z. Polym., 134, 149 (1953).
14. S.M. Ohlberg and S.S. Fenstermaker, J.Polym.Sci., 32, 514 (1958).
15. K.-H. Illers, Z. Electrochem., 70, 353 (1966).
16. J.R. McLoughlin and A.V. Tobolsky, J.Colloid Sci, 7, 555 (1952).
17. A.J. Kovacs, R.A. Stratton and J.D. Ferry, J.Phys.Chem., 67, 152 (1963).
18. K.-H. Illers, Makromol. Chem., 127, 1 (1969).
19. N.G. McCrum, B.E. Read and G. Williams, in "Anelastic and Dielectric Effects in Polymeric Solids", Wiley, London 1967, Chapter 6.
20. K.-H. Illers, Kolloid-Z. Z. Polym., 231,622 (1969).
21. M. Shen and E.H. Cirlin, J.Macromol.Sci., Phys, B(4), 947 (1970).
22. J.M. Crissman and E. Passaglia, J.Appl. Phys, 42, 4636 (1971).
23. Y.S. Papir and E. Baer, Mat.Sci.Eng., 8, 310 (1971).
24. M. Kakizaki, Y. Morita, K. Tsuge and T. Hideshima, Reports on Progress in Polymer Physics in Japan, 10, 397 (1967).
25. R. Simha and R.F. Boyer, J. Chem. Phys., 37,1003 (1962).
26. E.A.W. Hoff, D.W. Robinson and A.H. Willbourn, J.Polym.Sci., 18, 161 (1955).
27. J. Heijboer, P. Dekking and A.J. Staverman. In V.G.W. Harrison (Ed.) Proc. 2nd.Internl.Congr.Rheology, Academic Press, New York, P.123, (1954).
28. R.A. Haldon and R. Simha, J.Appl. Phys., 39, 1890 (1968).
29. R.A. Haldon and R. Simha, Macromolecules, 1, 340 (1968).
30. P.N. Lowell and N.G. McCrum, J.Polym.Sci., B, 9, 477 (1971).
31. J.M. Hutchinson and N.G. McCrum, Nature Physical Science, 236, 115 (1972).
32. N.G. McCrum, J.Polym. Sci., 54, 561 (1961).
33. L. Bohn, Kolloid-Z. Z. Polym., 201, 20 (1965).

A NOTE ON DIE SWELL

Yu Chen

Rutgers University

New Brunswick, New Jersey

ABSTRACT

In the extrusion of molten plastics through dies,
it is a common observation that the diameter of a
fibre or the thickness of a sheet is larger than that
of the die. The phenomenon is called die swell.

In this note, jets of three different types of
fluids are analyzed to give estimates of the contraction or the expansion in dimensions that takes place
downstream from the exit of the die. These three
different types of fluids are inviscid fluid,
Newtonian fluid, and a fluid with a certain kind of
viscoelasticity. Analysis show that contraction takes
place in the inviscid fluid. For Newtonian fluid,
contraction takes place at high Reynolds number. In
the case of a certain kind viscoelastic fluid, due to
the existence of normal stress, the jet is found to
expand its size downstream from the exit of the die.

INTRODUCTION

Some liquids of polymeric derivation possess
elastic properties which exhibit themselves in
phenomena such as elastic recoil, die swell, etc.
These liquids are sometimes called elastic liquids.
Perhaps a more appropriate name would be viscoelastic
fluids because they not only possess elasticity but
also viscosity. Flow behaviors of viscoelastic fluids

239

are much more complicated to analyze than the
classical fluids because of the nonlinearities that
exist in their constitutive descriptions. However,
it is useful in understanding some of the manifestations of the constitutive behaviors of this kind of
fluid on its global behavior in flow phenomena. One
such aspect of the manifestations is the so-called
die swell phenomenon.

It was said [1] that Newtonian fluids show a
small increase (about 10%) in cross-section radius
on emerging from the capillary at low Reynolds
number (Re<<1). For some viscoelastic fluids such
as polymer melts, a far greater increase takes place,
often as high as 100-200%. This is the phenomenon
known as die swell. In the extrusion processes of
the man-made fibre and plastic industries, diameter
increases of the order of 10% occur in rayon spinning
[2], and larger increases occur in the extrusion of
plastic melts through dies. In the formation of
plastic sheets by extrusion through a die with a
long thin slot as its cross-section, the emerging
plastic sheet is thicker than the slit.

The effect of expansion in dimension is so large
that it cannot be explained by the changing of volume
of the fluid. The compressibility of fluid is so
small that volume change of appreciable amount
requires enormous amount of pressure. Also,
observations of extruding experiments show that the
velocity of the fluid at the expanded sections is
smaller than that at the exit section of the die, a
further evidence that the swell is not due to change
of volume.

Free jet analysis in classical fluid mechanics
requires much sophisticated mathematics. Analysis
of jets of viscoelastic fluid have not appeared in
the literature due large part to the nonlinearities
existing in such problems. In the following, analysis
by conservation principles will be performed on jets
of inviscid fluid, Newtonian fluid, and a viscoelastic
fluid with an objective of estimating the possible
changes of dimensions.

Inviscid Fluids

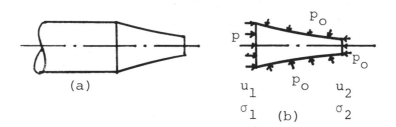

Fig. 1   A Jet of Inviscid Fluid

Figure 1(a) shows a inviscid fluid emerging with a velocity u_1 from a straight nozzle with cross-section σ_1. Figure 1(b) shows the jet isolated. The flow is assumed steady. Pressure at the exit is denoted by p_1, the ambient pressure is p_0. The constant section of the jet is labeled with velocity u_2 and area σ_2. From the consideration of continuity of flow we have

$$\rho\, \sigma_1\, u_1 = \rho\, \sigma_2\, u_2 \,, \tag{1}$$

the fluid is incompressible with density ρ. Therefore,

$$\frac{\sigma_2}{\sigma_1} = \frac{u_1}{u_2} \tag{2}$$

The ratio of the areas cannot be determined unless the ratio of the velocities is known. To achieve this we write the equation for momentum balance, viz.

$$(p - p_0)\sigma_1 = (\rho\sigma_2 u_2)u_2 - (\rho\sigma_1 u_1)u_1 \,. \tag{3}$$

The above equation involves a new unknown p, the pressure at the exit. However, since the flow is assumed to be steady, Bernoullis equation applies to each streamline; namely,

$$p + \frac{1}{2}\rho u_1^2 = p_0 + \frac{1}{2}\rho u_2^2 \,, \tag{4}$$

thus,
$$(p - p_o) = \frac{1}{2}\rho(u_2{}^2 - u_1{}^2) \ . \tag{5}$$

Substituting (2) and (5) into (3) and reducing yields

$$\frac{\sigma_2}{\sigma_1} = \frac{1}{2} \tag{6}$$

If the nozzle is circular in shape (6) can be written in terms of the radii of the two cross-sections.

Hence,
$$\frac{R_2}{R_1} = .707 \tag{7}$$

The ratio given in (6) is the classical result connected with the problem known as Borda's mouthpiece. [3] This ratio is called the coefficient of contraction in the literature of fluid mechanics.

Since the flow of inviscid fluid is devoid of effects of viscous forces, the contraction phenomenon is strictly due to inertial effects. The excess in pressure at the exit over the ambient gives rise to acceleration which in turn causes contraction.

Newtonian Fluids

Figure 1(a) this time will be used to illustrate the flow of a Newtonian fluid out of a circular pipe into the atmosphere. It is well known that the velocity distribution in this case is parabolic and given by

$$u_1 = \alpha(R_1{}^2 - r^2) \ , \tag{8}$$

where R_1 is the inner radius of the pipe, r is the radius vector, and α a constant depending on the pressure gradient in the flow and the viscosity of the fluid. Continuity of the flow requires that

$$2\pi \int_0^{R_1} \alpha(R^2 - r^2)r\,dr = \pi R_2{}^2\, u_2 \tag{9}$$

R_2 being the radius of the jet and u_2 the constant velocity of the jet. The velocity becomes constant by virtue of the fact the streamlines straighten themselves out sooner or later.

Performing the integration in (9) yields

$$\frac{\alpha}{2} R_1^4 = R_2^2 u_2 \quad . \tag{10}$$

The momentum balance for the jet can be written as follows.

$$2\pi \int_0^{R_1} (\rho u_1^2) r \, dr + 2\pi \int_0^{R_1} (p-p_o) r \, dr = \rho (\pi R_2^2) u_2^2 \tag{11}$$

The energy balance can be written by introducing a friction term to include the loss due to shearing deformation in the flow. Thus if we let L be the energy loss (or gain) a fluid particle along a streamline from the exit of the pipe to the end of the jet, then the energy balance equation states that

$$\frac{1}{2}\rho u_1^2 + p_1 = \frac{1}{2}\rho u_2^2 + p_o + L, \tag{12}$$

or,

$$p_1 - p_o = \frac{1}{2}\rho (u_2^2 - u_1^2) + L. \tag{13}$$

Although L is a constant for each streamline it is a function of the radius vector r. It is noted that the exact energy relationships are complicated since they must come from the integrated equations of motion of the field.

To solve (10) and (11) jointly for the area relationship we shall first assume that the integral in (11) containing the pressure difference term is small and can be neglected in the whole expression. Integrating (11) after dropping the term yields

$$\frac{\alpha^2}{3} R_1^6 = R_2^2 u_2^2 \quad . \tag{14}$$

From (10) and (14) we obtain

$$\frac{R_2}{R_1} = .866 ,$$

thus,

$$\frac{\sigma_2}{\sigma_1} = (.866)^2 = .750.$$

The above ratio of the radii was also given in [4]. The dropping of the pressure term implies that the flow rate is so high that at the exit the total momentum is due to the velocity. Therefore this situation can be categorized under high Reynolds number flow.

Otherwise, we shall simply substitute (13) into (11) and integrate the whole expression. Such a calculation is meaningless unless the exact variation of L is known. Therefore we conclude that the global type of analysis of jet which give exact solution for the contraction ratio in the case of inviscid fluid is found to be less productive here in the Newtonian fluid. Nevertheless, some relevant information was extracted.

A Viscoelastic Fluid

A similar approach as that used in the previous two cases will be applied to a jet of viscous fluid exhibiting elastic behavior. For the flow of this type of fluid the velocity profile is usually less parabolic and tends to be uniform except at the high shear region near the wall. To simplify calculation we shall assume a trapizoidal velocity profile, the linear portion of the velocity profile is restricted to the region $R>r>\xi R$, where ξ is a number less than unity. Figure 2 shows the jet with its associated velocity profiles at the ends.

Let the normal stress in the axial direction be τ_{zz} which is assumed to be proportional to the velocity gradient.

Thus,

$$\tau_{zz} = k \left(\frac{\partial u_1}{\partial r}\right)^2 \tag{15}$$

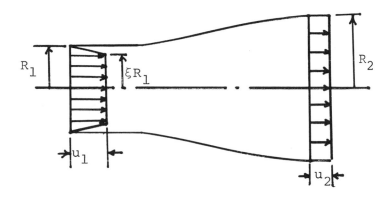

Fig. 2   A Jet of a Viscoelastic Fluid

From the velocity profile we have

$$\frac{\partial u_1}{\partial r} = \frac{u_1}{(1-\xi)R_1} \quad . \tag{16}$$

Therefore,

$$\tau_{zz} = k \frac{u_1^2}{(1-\xi)^2 R_1^2} \quad . \tag{17}$$

The continuity equation is given by

$$\pi(\xi R_1)^2 u_1 + \pi[(1-\xi^2)R^2]\frac{u_1}{2} = (\pi R_2^2)u_2 , \tag{18}$$

or

$$u_2 = \frac{1}{2}(1+\xi^2)\left(\frac{R_1}{R_2}\right)^2 u_1 \tag{19}$$

The momentum balance can be written as

$$2\pi \int_0^{R_1} (\rho u_1^2 - \tau_{zz}) r\,dr = \pi\rho R_2^2 u_2^2 \tag{20}$$

Since the velocity profile is deliberately chosen to be simple, the above integral can be evaluated as follows,

$$\pi(\xi R_1)^2 \rho u_1^2 + \frac{\pi}{4}(1 - \xi^2) R_1^2 \rho u_1^2 ,$$

$$-k\pi u_1^2 = \pi\rho R_2^2 u_2^2$$

or

$$\frac{R_1^2}{4}(1 + 3\xi^2)\rho u_1^2 - k u_1^2 = R_2^2 \rho u_2^2 . \tag{21}$$

Substituting (19) into (21) and reduce, we have

$$(1 + \xi^2)\left(\frac{R_1}{R_2}\right)^4 - (1 + 3\xi^2)\left(\frac{R_1}{R_2}\right)^2 + \frac{4k}{\rho} = 0 . \tag{22}$$

Solving,

$$\left(\frac{R_1}{R_2}\right)^2 = \frac{1 + 3\xi^2 \pm \sqrt{(1 + 3\xi^2)^2 - \frac{16k}{\rho}(1 + \xi^2)^2}}{2(1 + \xi^2)^2} \tag{23}$$

Mathematically speaking, if k is zero we would have

$$\left(\frac{R_1}{R_2}\right)^2 = \frac{1 + 3\xi^2}{(1 + \xi^2)^2} ,$$

which for $\xi < 1$ always gives a ratio greater than unity. Therefore without the normal stress the jet always contracts. On the other hand, for $k > 0$ it can be observed that the ratio R_1/R_2 will be greater than unity. Therefore the die swell phenomenon is seen to be associated with the elastic property of the fluid as is depicted by the constant k in this particular case.

References

1.   J. R. A. Pearson, "Mechanical Principles of
     Polymer Melt Processing," Pergamon Press, 1966,
     pp. 48-52.

2.   A. S. Lodge, "Elastic Liquids," Academic Press,
     1964, pp. 242-244.

3.   L. M. Milne-Thomson, "Theoretical Hydrodynamics,"
     second edition, MacMillan Company, New York,
     1950, pp. 20-21.

4.   A. G. Frederickson, "Principles and Applications
     of Rheology," Prentice-Hall Inc., Englewood
     Cliffs, N.J. 1964, p. 218.

THE EFFECTS OF MOLECULAR STRUCTURE ON THE MELT RHEOLOGY OF LOW DENSITY POLYETHYLENE

Chan I. Chung, Jack C. Clark* and Lowell Westerman

Esso Research and Engineering Company

Baytown, Texas

ABSTRACT

A rapid method for obtaining information on average molecular weights, molecular weight distribution, molecular size, and long chain branching (LCB) characteristics of low density polyethylene (LDPE) from gel-permeation chromatography (GPC) and intrinsic viscosity (IV) was recently reported. This investigation is primarily concerned with the effect of molecular weight and molecular structure, obtained by the use of the GPC-IV method, on the melt rheology of whole LDPE. Some of the assumptions involved in the GPC-IV method are also examined. It is found that the melt viscosity of whole LDPE depends not only on weight average molecular weight, but also strongly on the LCB characteristics. The combined influence of these two molecular parameters on melt viscosity can be described through the effect which each has on the weight-average mean square radius of gyration of the polymer coil, $(\overline{S^2})_w$. The experimental data indicate that the dependence of zero shear viscosity on $(\overline{S^2})_w$ is substantially greater for LDPE than for linear polymers. The apparent flow activation energy at zero shear is found to be about 12 kcal/mole for all whole LDPE samples of differing LCB characteristics studied in this investigation. Our calculations based on various assumptions in the GPC-IV method suggest that the polydisperse model of the Zimm and Stockmayer equation, which relates molecular weight and branching frequency to the branching

* Present address: Foster Grant, Inc.
                    Leominster, Mass.   01453

parameter (g), and the exponent of b = 1/2 in the equation, g^b = $[\eta]_B/[\eta]_L$, are the best choices for whole LDPE. $[\eta]_B$ is the IV of branched molecule and $[\eta]_L$ is the IV of linear molecule of the same molecular weight.

I. INTRODUCTION

The dependence of melt viscosity on the molecular weight has been well established for most linear polymers (1-11). The effects of short chain branching (SCB) and long chain branching (LCB) in polymers on the melt viscosity have been also studied by many investigators (12-21). Low density polyethylenes (LDPE) manufactured by high pressure, free radical processes normally have very complicated branch structures involving both SCB and LCB. Over the last several years a few investigators (22,22), primarily working with fractions, have developed an efficient method of measuring the LCB in LDPE as well as other molecular parameters such as molecular weight (MW) and molecular weight distribution (MWD) by the combined use of gel permeation chromatography (GPC) and intrinsic viscosity (IV). This method will be referred to in this paper as the GPC-IV method. Wild, Ranganath and Ryle (21) developed a method of obtaining the molecular parameters of whole LDPE from the molecular parameters of its individual fractions. Their method requires fractionation and the characterization of each fraction. Fractionation is a very time consuming and tedious process. Furthermore, the MWD in most commercial LDPE is broad (M_w/M_n >12) and the molecular parameters obtained for their fractions cannot be easily translated into the molecular parameters of the parent whole polymer.

Drott and Mendelson (22) proposed a method of obtaining the molecular parameters for whole LDPE without the need of fractionation, as well as for LDPE fractions. Mendelson, et al. (20) studied the effect of LCB as measured by the GPC-IV method on the melt rheological properties of LDPE fractions and serveral experimental whole LDPE samples. Their study shows that the melt viscosity of LDPE, for both fractions and whole polymers, depends not only on the MW but also strongly on the LCB. Their experimental whole LDPE samples, however, have very narrow MWD (M_w/M_n <2) approaching that found for fractions, and thus are not representative of commercial whole polymers. It is desirable to investigate the relationship between the molecular parameters and the melt rheology of whole LDPE with broad MWD. We have measured the molecular parameters of several commercial and experimental whole LDPE samples of broad MWD by the use of the GPC-IV method which is essentially identical to the method

proposed by Drott and Mendelson. Our method differs from theirs
only in the averaging process to obtain the average molecular
size.

   High polymer melts are categorized as pseudoplastic fluids
since their viscosities, unlike Newtonian fluids, decrease as the
shear rate is increased. However, there is a region of very low
shear rates where the viscosity does not decrease with increasing
shear rate. This shear rate region is called the low shear
Newtonian region and the corresponding shear rate independent
viscosity is called the low shear Newtonian viscosity or zero
shear viscosity. The zero shear viscosity is usually found at
very low shear rates below 0.005 sec^{-1} for LDPE as well as for
HDPE. Zero shear viscosity forms the basis for theoretical
treatments of the effects of molecular structures on melt flow
behavior (1-3, 23). Zero shear viscosity data are therefore
important in an investigation directed toward obtaining a better
understanding of how molecular parameters influence melt flow
behavior. We have measured zero shear viscosities of whole LDPE
samples at several temperatures using a Weissenberg Rheogoniometer.
This paper describes the relationship between the molecular
parameters and the melt rheology at very low shear rates for
whole LDPE. This paper also examines some of the assumptions
involved in the GPC-IV method. High density polyethylene (HDPE)
was studies to provide a point of reference and these results
are also presented.

II. MEASUREMENT OF MOLECULAR WEIGHT AND LONG CHAIN BRANCHING
     IN WHOLE LDPE

A. Background

   The GPC-IV method for determining the molecular parameters
(MW, MWD, molecular size, and LCB) of LDPE is based on the
ability to calculate molecular sizes from a combination of GPC
and IV data, and on the work of Zimm and Stockmayer (24) which
describes the effect of LCB on molecular size. This method
has been described previously by Drott and Mendelson (22). The
details of this method are examined in this paper to point out
some of the problems associated with its use. GPC is commonly
employed to rapidly characterize the molecular parameters of
polymers. Benoit (25) observed that the separation on the GPC
column appeared to be primarily controlled by the hydrodynamic
volume (h^3) of the polymer molecules. He and coworkers have
found that a universal calibration curve (Figure 1) can relate
the elution volume to the hydrodynamic·volume, as given by the
product of MW and IV, for a number of linear and branched

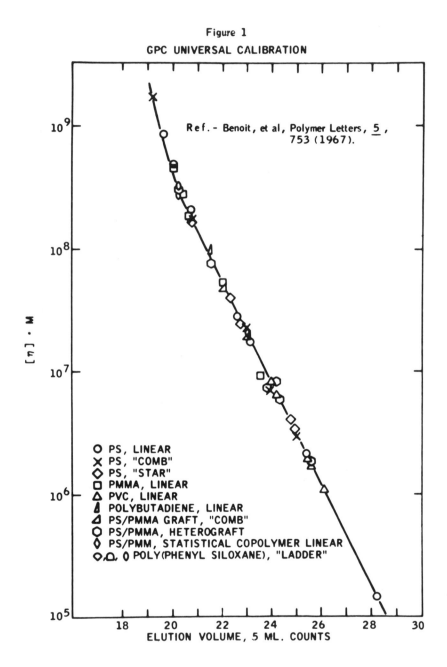

Figure 1
GPC UNIVERSAL CALIBRATION

polymers (26). For linear polymers IV depends on MW according
to the Mark-Houwink equation (27) shown below and thus the GPC
elution volume can be directly converted into the MW.

$$[\eta]_L = K M_L^\alpha \tag{1}$$

Figure 2 shows a typical GPC calibration curve and trace. Any
average of the MW can be calculated from Figure 2. However, the
simple treatment of the GPC data outlined above for linear
polymers cannot be applied to branched polymers since the IV of
branched polymers is not simply a function of MW but also depends
on LCB characteristics.

The radius of gyration of the molecule with LCB is less than
that of its linear analog of the same MW. The ratio of the mean-
square radii of gyration for branched and linear molecules of the
same MW is commonly expressed by the branching parameter, g,
(24, 28).

$$g = \overline{s_B^2} / \overline{s_L^2} \tag{2}$$

The value of g depends on the LCB characteristics such as the
functional nature of branch points, the number of branch points
in a molecule, the length of branches, and the distribution of
branches among the molecules. Zimm and Stockmayer (24) have
derived a series of equations relating g and various LCB
characteristics for both monodisperse and polydisperse systems.
The equation for a polydisperse system with randomly distributed
trifunctional LCB is

Figure 2
TYPICAL GPC CALIBRATION AND GPC TRACE

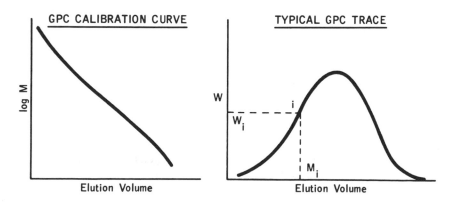

$$g_w = \frac{6}{n_w} \left[\frac{1}{2} \frac{(2+n_w)^{1/2}}{(n_w)^{1/2}} \cdot \ln \frac{(2+n_w)^{1/2}+(n_w)^{1/2}}{(2+n_w)^{1/2}-(n_w)^{1/2}} -1 \right] \tag{3}$$

g_w and n_w are, respectively, the weight average branching parameter and weight average number of branch points per molecule.
Assuming that the LCB frequency defined by $\lambda = n/M$ is essentially
constant for all MW species in a given sample, equation 3 may
be rewritten in terms of λ and M as,

$$g_w = \frac{6}{\lambda M} \left[\frac{1}{2} \frac{(2+\lambda M)^{1/2}}{(\lambda M)^{1/2}} \cdot \ln \frac{(2+\lambda M)^{1/2}+(\lambda M)^{1/2}}{(2+\lambda M)^{1/2}-(\lambda M)^{1/2}} -1 \right] \tag{4}$$

$$= g_w (\lambda M)$$

Intrinsic viscosity has been related to the radius of gyration
by Flory and Fox (29) as follows,

$$[\eta] = \phi' \frac{(\overline{s^2})^{3/2}}{M} \tag{5}$$

where ϕ' is a universal constant. It can be shown from equations
2 and 5 that the ratio of the IV of a branched molecule to that
of its linear analog of the same MW should be related to g
according to equation 6.

$$g^{3/2} = [\eta]_B/[\eta]_L \tag{6}$$

There is now a considerable amount of evidence suggesting that
equation 6 exaggerates the effect of branching on IV (30-37).
Stockmayer and Fixman (28) suggested that IV would depend on
effective hydrodynamic radius (h) rather than on the radius of
gyration employed in equation 5 as given by,

$$[\eta] = \phi \frac{\overline{h^3}}{M} \tag{7}$$

where ϕ is taken as a universal constant applicable to both
linear and branched molecules. They explained the failure of
equation 6 to accurately predict the influence of branching on
IV by the difference in the effects of branching on hydrodynamic
radius and on radius of gyration. Zimm and Kilb (32) proposed
1/2 as more appropriate exponent than 3/2 in relating IV to
branching in equation 6.

$$g^{1/2} = [\eta]_B/[\eta]_L \tag{8}$$

Although numerous experimental results (20,21,30-37) support the
use of equation 8, there is no strong theoretical justification
for the exponent of 1/2 and the validity of equation 8 is still
open to question. A more detailed review of this subject was

recently published by Graessley (38).

By combining equations 1 and 8, we obtain

$$[\eta]_B = K M_B^{\alpha} \, g^{1/2} \tag{9}$$

If linear and branched molecules are characterized by the same GPC elution volume (i.e., the same hydrodynamic volume), it follows from equation 7 that the products of their IV and MW can be equated.

$$M_L [\eta]_L = M_B [\eta]_B \tag{10}$$

B. Molecular Weight and Long Chain Branching in Whole LDPE

For a truly monodisperse, branched polymer the MW can be calculated from the GPC elution volume (peak value) and the experimentally determined IV by using equation 10 and the GPC universal calibration curve. Then the IV of a linear polymer of the same MW is calculated by using the Mark–Houwink equation (equation 1). The value of g is calculated from the ratio of the experimentally determined IV of the branched polymer to that calculated for the linear polymer of the same MW by using equation 8. λ (or n) can now be calculated from the values of g and M by using the appropriate Zimm and Stockmayer equation. The analysis of the GPC-IV data for a polydisperse, branched polymer is not so straightforward since the IV of the polymer at each GPC elution volume is not known.

The IV of a polydisperse polymer can be calculated from the weight fractions (w) and IVs of all its fractions as follows.

$$[\eta] = \sum_i^N W_i [\eta]_i \tag{11}$$

By combining equations 9 and 11, we obtain

$$[\eta]_B = K \sum_i^N W_i M_{Bi}^{\alpha} \, g_i^{1/2} \tag{12}$$

Each GPC elution volume cannot be considered as a truly mono-disperse system and thus it would be more appropriate to use the Zimm and Stockmayer equation for a polydisperse system in relating g with λ and M. By combining equations 1, 4, 9 and 10, we obtain the following equation for each GPC elution volume.

$$M_{Li}^{1+\alpha} = M_{Bi}^{1+\alpha} \, [g_w(\lambda \, M_{Bi})]^{1/2} \tag{13}$$

The substituting of equation 4 into equation 12 gives

$$[\eta]_B = K \sum_i^N W_i M_{Bi}^\alpha [g_w(\lambda M_{Bi})]^{1/2} \tag{14}$$

The values of λ and M_{Bi}'s can now be obtained by an iterative process by using equations 13 and 14. The necessary experimental data are the GPC calibration curve for the linear polymer, the IV and GPC trace of the whole branched polymer. First, M_{Bi} and g_{wi} of each GPC elution volume are calculated for an assumed λ by using equation 13 and the GPC calibration curve for the linear polymer. Than the IV of the whole polymer is calculated from equation 14. If the calculated IV of the whole polymer does not agree with the experimentally measured value, λ is varied until they agree within a preset limit. Once the correct λ, g_{wi}'s and M_{Bi}'s are obtained, the various MW averages of the whole polymer such as number-average MW (M_n), and weight-average MW (M_w), can be easily calculated. It is also possible to calculate various average values for the mean square radius of gyration. This is accomplished by computing the product of the MW (M) and the branching parameter (g) of each GPC elution volume and applying an appropriate averaging procedure. For example, the weight average value, $(\overline{M \cdot g})_w = \sum_i^N W_i M_i g_i$ may be calculated. This parameter is proportional to the weight average value of the mean square radius of gyration as shown below,

$$(\overline{M \cdot g})_w = b \; (\overline{s^2})_w \tag{15}$$

where b is a constant. This equation applies to the case of linear polymers when g has the value of unity.

Although the GPC-IV method has unresolved problems associated with it, it is considered to be the best method currently available to obtain the molecular parameters of whole LDPE.

III. EXPERIMENTAL

A. Intrinsic Viscosity

The IVs of the polyethylene samples were measured at 135°C using decalin as the solvent in modified Ubbelohde viscometers.

The Mark-Houwink equation derived for linear polyethylene by Chiang (39) was used to relate the IV to the viscosity average MW:

$$[\eta]^{135°C}_{Decalin} = 6.2 \times 10^{-4} \ \bar{M}_v^{0.70} \tag{16}$$

The Mark-Houwink exponent, $\alpha = 0.7$, in equation 16 is strictly applicable only to linear polyethylene in decalin at 135°C; however, in the computations described in section II, α is assumed to have the same value in the GPC solvent as in decalin.

B. Gel Permeation Chromatography

A Waters' Model 100, Gel Permeation Chromatograph equipped with polystyrene-gel columns having nominal permeability limits of 10^7, 10^5, 10^4, and 250A° was employed. The instrument was operated at 135°C. Trichlorobenzene stabilized with 2,6-ditertiary butyl phenol was used as the solvent at a flow rate of one milliliter per minute.

Samples were prepared for injection at a concentration of 0.25 wt.%. An injection time of one minute was employed to give a column loading of 0.0025 gram.

Calibration of elution volume in terms of the MW of linear polyethylene was accomplished by the use of fractions of linear polyethylene, hydrogenated polybutadiene samples and several whole HDPE samples having broad MWD and IV from 1.0 to 2.35 dl/g. After an initial calibration curve was obtained with the linear polyethylene fractions, slight adjustments were made, particularly to the high molecular weight portion of the calibration, until the IV calculated from the GPC traces on the whole HDPE samples agreed with measured values to within 10%.

C. Zero Shear Viscosity

The zero shear viscosities of all samples were measured using a Model D Weissenberg Rheogoniometer (WR). WR is basically a cone and plate viscometer (40) and it is capable of measuring viscosities at extremely low shear rates. The weight of the polymer to properly fill the gap between the cone and plate was determined and a pre-weighed, thin disk specimen were used to eliminate the necessity of opening the chamber after the specimen melted to remove excess material. This procedure greatly reduced the danger of specimen oxidation. Specimen oxidation gives erroneous viscosity. Viscosity measurements were made at several temperatures with a given specimen in the WR. The temperature of the specimen and the gap between the cone and the plate were very carefully adjusted at each temperature. All viscosities were measured in steady rotation.

IV. RESULTS AND DISCUSSION

Five commercial LDPE samples, one experimental LDPE sample
(LDPE-F) with a very broad MWD, and two commercial HDPE samples
were studied in this investigation. The intrinsic viscosities
and the molecular parameters of the samples obtained by using
the GPC-IV method are summarized in Table 1. Table 2 summarizes
the zero shear viscosities of the sample at several temperatures
measured by WR. Viscosity measurement by WR requires high
degree of skill and care in comparison to Instron Capillary
Rheometer. The reliability of our WR viscosity data is
demonstrated in Figure 3 which shows an excellent agreement
between WR data and Instron Capillary Rheometer data.

A. The Effect of Long Chain Branching on Melt Viscosity

It is well established that an expression in the form of
equation 17 adequately describes the relationship between zero
shear viscosity (η_o) and weight average molecular weight ($\overline{M}_w$)
for linear polymers (1-11).

$$\eta_o = K' \overline{M}_w{}^a \tag{17}$$

where K' and a are constants. Figure 4 demonstrates that an
equation of this type is inadequate for LDPE. The poor
correlation indicates that η_o depends not only on $\overline{M}_w$ but also on
some other molecular feature.

Theoretical and empirical work of Bueche (1-3), Fox and
Allen (23) and others leads to the use of mean square radius of
gyration, rather than MW for describing the melt viscosity
behavior on a molecular basis. Fox and Allen (23) have proposed
a universal parameter for use in correlating zero shear viscosity
behavior of polymers of different structure. This parameter, X,
is defined as follows:

$$X = [(\overline{S_o^2})/M] (Z_w/V) \tag{18}$$

where $(\overline{S_o^2})$ is the unperturbed mean square radius of gyration of
a polymer molecule of molecular weight, M, having a weight-
average number of chain atoms, Z_w, and V is the specific volume
of the melt at the temperature specified. Accordingly, the
isothermal zero shear viscosity data for a polymer may be
expressed by the relationship:

$$\eta_o = K_t (\overline{S_o^2})_w{}^a \tag{19}$$

Table 1

The Molecular Parameters Determined by the GPC-IV Method

Sample	$[\eta] \frac{135°C}{Decalin}$	$\bar{M}_n \times 10^{-3}$	$M_w \times 10^{-3}$	$M_z \times 10^{-3}$	$\frac{\bar{M}_w}{\bar{M}_n}$	$(\bar{M}g)_n \times 10^{-3}$	$(\bar{M}g)_w \times 10^{-3}$	$(\bar{M}g)_z \times 10^{-3}$	$\frac{(\bar{M}g)_w}{(\bar{M}g)_n}$	$\lambda \times 10^4$
LDPE-A	0.80	9.1	148	1669	16.3	7.4	21.3	38.1	2.9	1.7
-B	0.89	10.4	239	2007	23.0	8.0	22.4	38.9	2.8	2.03
-C	0.99	14.2	323	2244	22.7	9.89	24	38.8	2.4	2.27
-D	1.05	16.4	199	1218	12.1	12.1	30.2	50.2	2.5	1.33
-E	1.12	15.5	186	915	12.0	12.2	34.2	55.6	2.8	1.14
-F	1.63	20.7	1303	5581	62.9	13.5	36.9	54.7	2.7	2.28
HDPE-A	1.50	7.2	88.8	706	12.3	–	–	–	–	0
-B	2.38	8.4	194	1055	23.1	–	–	–	–	0

The molecular parameters were claculated with the following assumptions:
1. $g^{1/2} = [\eta]_B/[\eta]_L$
2. polydisperse in molecular weight
3. trifunctional branches
4. random distribution of branches among the molecules

Table 2

The Zero Shear Viscosities Measured by Weissenberg Rheogoniometer

Sample	Zero shear viscosity x 10^{-3}, poise					ϕApparent flow activation energy at Zero shear, Kcal/mole
	130°C	150°C	170°C	190°C	210°C	
LDPE-A	40.6	18.3	10.2	5.6	3.3	12.0
-B	47.3	23.2	11.7	6.8	-	12.0
-C	108.2	52.1	26.9	15.5	9.7	11.7
-D	431.6	196.4	101.2	53.1	33.5	12.4
-E	2185	1019	476	268	140	13.2
-F*	3435	1586	793	423	242	12.8
HDPE-A	-	99.6	71.2	50.7	-	6.5
-B*	-	1991	1445	1031	-	6.4

*The viscosities of LDPE-F and HDPE-B slowly increased down to the minimum experimental shear rate of 0.0017 sec⁻¹. The viscosities at 0.0017 sec⁻¹ are assumed to represent zero shear viscosities.

ϕ Apparent flow activation energy was obtained by linear regression analysis.

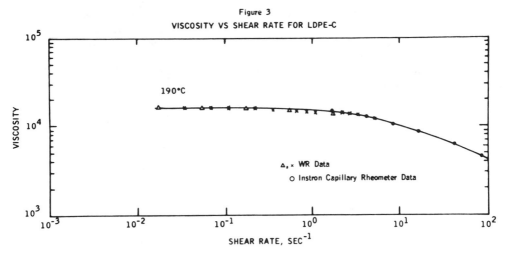

Figure 3
VISCOSITY VS SHEAR RATE FOR LDPE-C

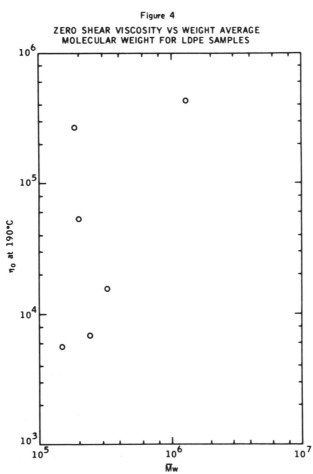

Figure 4
ZERO SHEAR VISCOSITY VS WEIGHT AVERAGE
MOLECULAR WEIGHT FOR LDPE SAMPLES

where K_t is a constant dependent on the specific structure of
the polymer and temperatures, a = 1 when M is less than the
critical molecular weight for entanglement, M_c, and a = 3.5 for
$M > 2M_c$. The weight-average mean square radius of gyration,
$(\overline{S^2})_w^0$, is suggested to be more generally appropriate for use
in describing melt viscosity behavior than is M_w.

   According to equation 15, the weight-average value of $(\overline{S^2})$
for LDPE can be derived from the data obtained by the GPC-IV
method. Since $(\overline{S^2})_w$ is proportional to $(\overline{Mg})_w$, we should expect
$(\overline{Mg})_w$ to describe the viscosity behavior of LDPE in accordance
with the following equations:

$$\eta_o = K_t' \ (\overline{Mg})_w^{\ a} \tag{20}$$

or

$$\ell n \ \eta_o = A + a \ \ell n \ (\overline{Mg})_w \tag{21}$$

Figure 5 shows the relationship described by equation 17 (or
equation 21) between η_o and M_w (note that $(\overline{Mg})_w = M_w$ for linear
polymers since g = 1) for the two HDPE samples at 3 tempera-
tures; the slope, a, is about 3.8 at all temperatures. This is
in good agreement with the reported values around 3.5 for linear
polymers, considering that the slope is based on only 2 data
points. Figure 6 shows that equation 21 can indeed describe the
relationship between η_o and $(\overline{Mg})_w$ for the six LDPE samples at 5
different temperatures. The data of the LDPE-D sample deviate
noticeably from the line. The reason for such deviation is not
understood. The constants (A and a) in equation 21 were
obtained at each temperature by linear regression analysis, and
they are listed in Table 3. We find that the slope for LDPE is
about 8 at all temperatures. This is abnormally high in contrast
to the slope of about 3.5 predicted by Bueche's theory and also
proved for many linear polymers (3-11). Mendelson, et al. (20)
also found a very high slope of about 6.6 for their LDPE
fractions and whole LDPE samples with a very narrow MWD.

   The extreme sensitivity of η_o to a change in $(\overline{Mg})_w$ means
that small variations in MW or in LCB characteristics can cause
much larger variations in viscosity than are experienced with
linear polymers. Good measurements of η_o, therefore, should be a
quite sensitive way to detect this type of variation in LDPE
which appear otherwise to be identical.

   The cause of the higher exponent in the Bueche equation (3)
for LDPE is not clear. Kraus and Gruver (19) have noted for
trichain and tetrachain polybutadienes that high exponents are
obtained when the branch length is above some critical length.

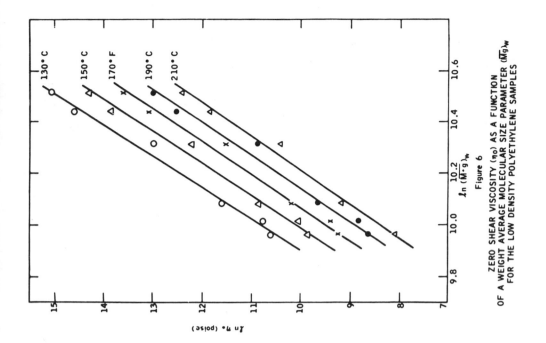

Figure 6

ZERO SHEAR VISCOSITY (η_0) AS A FUNCTION
OF A WEIGHT AVERAGE MOLECULAR SIZE PARAMETER ($\overline{M}\overline{g}_w$)
FOR THE LOW DENSITY POLYETHYLENE SAMPLES

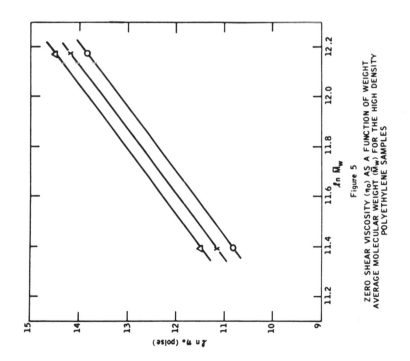

Figure 5

ZERO SHEAR VISCOSITY (η_0) AS A FUNCTION OF WEIGHT
AVERAGE MOLECULAR WEIGHT ($\overline{M}_w$) FOR THE HIGH DENSITY
POLYETHYLENE SAMPLES

Table 3

The Regression Constants in the Relationship, $\ln\eta_o = A + a\ln(\overline{Mg})_w$, for Low Density Polyethylene Samples

Temperature	A	a	Correlation Coefficient
130°C	−71.61116	8.23690	0.993
150°C	−71.67513	8.16921	0.993
170°C	−70.86478	8.02352	0.993
190°C	−70.48167	7.92832	0.992
210°C	−68.01186	7.63791	0.993

They have proposed a critical branch length for entanglement and suggest that entanglement of the branches contributes to the much more rapid increase in η_o than is experienced with unbranched polymers or with branched polymers where the branch length is less than the critical value. Following this same line of reasoning, one may conclude that the branch length of all LDPE resins examined thus far are above the critical length for entanglement.

B.   The Effect of Long Chain Branching on Apparent Flow Activation Energy

The temperature coefficient of viscosity is defined by the familiar Arrhenius equation, below, where E is the apparent flow activation energy, R is the gas constant, T is the absolute

$$\eta = A \ e^{E/RT} \tag{22}$$

temperature, and A is a constant whose value depends on the specific polymer in question.

There has been a considerable amount of work published in the literature which attempts to establish how molecular weight and chain structure influence the value of E. It seems to be well established that the apparent flow activation energy at zero shear (E_o) increases fairly rapidly with M below M_c. This behavior is depicted in the figure below.

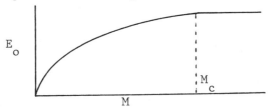

Data on normal paraffins and linear polyethylene indicate that the value of E_o above M_c should be about 5 Kcal/mole (41). Measured E_o values for various HDPEs are found to vary between 6 and 8 Kcal/mole (15,42). On the other hand, E_o values for LDPE are reported in the range of about 11 to 13 Kcal/mole. The reason for this large difference between E_o for LDPE and HDPE is not known with certainty. The cause has variously been attributed to the higher degree of short chain branching or to the presence of long chain branching in LDPE compared to HDPE.

In Figure 7, $\ln\eta_o$ is shown as a function of the inverse of absolute temperature for both HDPE and LDPE. Linear relationships between $\ln\eta_o$ and 1/T are found as is expected. The slopes of the lines were determined by linear regression analysis and the E_o calculated from the slopes are presented in Table 2. It is seen from Table 2 that E_o is about 12-13 Kcal/mole for all 6 samples where as it is about 6.5 Kcal/mole for the 2 HDPE samples. There is also an indication that E_o of LDPE increases slightly with $(\overline{Mg})_w$ or $(Mg)_z$.

Our LDPE samples had nearly identical SCB characteristics, but had LCB frequencies ranging from λ= 1 x 10^{-4} to 2.3 x 10^{-4} and widely different MWD. Thus, for those samples which we have examined thus far, E_o does not appear to be sensitive to variations in λ or MWD within the range studied.

Tung (15) observed a higher E for LDPE fractions than for HDPE fractions. He also studied a fraction of a linear ethylene-1-butene copolymer where the frequency of ethyl side groups approached the SCB frequency found in typical LDPE resins. His data show that E_o for this sample is not greatly different from samples of HDPE containing far less SCB. He concluded that SCB could not account for the difference in E_o between LDPE and HDPE.

Boghetich and Kratz (43) have reported activation energy data on a series of polyethylene samples which differed in LCB characteristics as determined by an infrared method. Their data suggest that E increases with LCB frequency. One of the samples of their series was available to us for measurement of λ by the

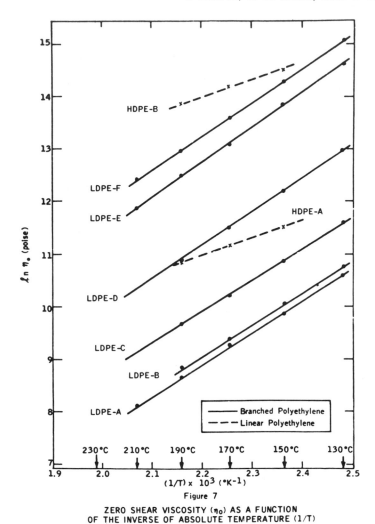

Figure 7

ZERO SHEAR VISCOSITY (η_0) AS A FUNCTION
OF THE INVERSE OF ABSOLUTE TEMPERATURE (1/T)

GPC-IV method. This sample had the highest LCB frequency and highest E value in their series. We found a value of $\lambda = 1.3$ x 10^{-4} for this sample, placing it within the range of the other typical LDPE resins which we have examined. The data of Boghetich and Kratz together with our results suggest an increase in E_0 with increasing λ below a value of 1 x 10^{-4} followed by a leveling out of E_0 at higher LCB frequencies. One might speculate, by analogy with the behavior of E_0 as a function of M, that E_0 rises rapidly from a value of 5 Kcal/mole with the introduction of a few long chain branches into the polyethylene molecules and then levels off at a value in the range of 11 to 13 Kcal/mole in the λ region above 1 x 10^{-4}. Alternately, E_0 may be a function of branch length until some critical branch length

is attained. More data are certainly required in the low LCB
frequency region to clarify this behavior.

It is worthy of note that commercial HDPE resins show a
rather wide variation in E_0 (42). This fact, together with
reports (44) that some HDPEs contain long chain branches, is
also suggestive of the involvement of LCB in increasing E_0 at
much lower levels than is normally found for commercial LDPE
resins.

C.   Assumptions in the GPC-IV Method

There are four major assumptions which are made in using
the GPC-IV method to evaluate average molecular weights, long
chain branching frequency, λ, and the size parameter, $(Mg)_w$.

The basic assumption which allows us to make use of the
GPC-IV method is concerned with the relationship between GPC
elution volume and the effective hydrodynamic volume of the
polymer molecules in solution. Benoit (25,26) has demonstrated
that polymers of different structure will have the same GPC
elution volume when their hydrodynamic volumes are the same.
According to Stockmayer and Fixman (28) the product of intrinsic
viscosity and molecular weight, $[\eta]\cdot M$, is directly proportional
to hydrodynamic volume. On this basis the equality given by
equation 10 will apply at any specified elution volume.

The essential validity of this assumption is demonstrated
by the data shown in Figure 1 where data for a wide variety of
polymer structures is plotted as log $[\eta]M$ versus elution volume.
The fact that the data form a universal calibration curve is
taken as good evidence for the validity of equation.

The second assumption involves the way in which the long
chain branching frequency, λ, varies with molecular weight
within any single whole LDPE sample. Since there is, at
present, no established method for determining how λ varies
with MW, we have assumed that λ is constant in any one whole
LDPE sample. The experiment of Mendelson and Drott (45)
supported this assumption. Very recently Wild, et al. (21)
found that λ was essentially constant in a LDPE sample with low
level of branching but increased with MW in the highly branched
LDPE samples. We have also studied the dependence of λ on MW
by fractionating whole LDPE samples. Our results show that λ
increases with MW but the assumption of a constant λ for a given
sample does not lead to significant error. Further discussion
on this subject will be presented elsewhere by Westerman and
Clark (46).

The third assumption involves the equations which relate
molecular weight and branching frequency to the branching

parameter, g. The equations describe how variations in the molecular weight and branching frequency influence the value of g. Three models have been provided by Zimm and Stockmayer (24) for the case of random trifunctional branching. These three models all consider a random distribution of branch units within one molecule with branches of random length. The first model (ZS-p), chosen in our initial work, considers a polydisperse collection of branched molecules where there is also a random distribution of branches among the molecules. The second model (ZS-m) considers a system which is monodisperse in molecular weight, but the distribution of branch units among the molecules is random. The third model (ZS-u) is also a monodisperse model, but differs from ZS-m in that it requires each molecule to have the same number of branch units. This case gives results which are essentially the same as those obtained with the ZS-m model, so we will limit our attention to results obtained by using the first two models.

The last assumption involves a choice of the exponent of g in the equation which relates g to the intrinsic viscosity ratio of a linear and branched polymer species of the same molecular weight. The literature (30-37) indicates that the value of this exponent may vary between 1/2 and 3/2 depending upon the nature and degree of branching. Our initial choice was 1/2.

The influence which the last two assumptions have on the results obtained are discussed below.

Calculations were made on two samples for which the weight-average molecular weight, $\overline{M}_w$, had been determined by light scattering. The GPC-IV method was employed with modifications to allow calculations of $\overline{M}_w$ for these samples using two ZS-models and three different values for the exponent of g. The results of these calculations are shown compared to the results from light-scattering in the table below.

Method for $\overline{M}_w$	$\overline{M}_w \times 10^{-3}$	
	LDPE-C	LDPE-E
Light-Scattering	278	179
$g^{1/2}$ (ZS-p)	323	186
(ZS-m)	266	178
$g^{0.7}$ (ZS-p)	320	197
(ZS-m)	282	184
$g^{3/2}$ (ZS-p)	714	200
(ZS-m)	367	210

It can be seen from this table that $\overline{M}_w$ values calculated with $g^{1/2}$ using either model agree fairly well with the light-scattering values, considering that $\overline{M}_w$ values determined by light-scattering are probably no more accurate than $\pm 20\%$ on unfractionated LDPE samples. The results with $g^{0.7}$ are also reasonable; however, if we go to $g^{3/2}$ we see major differences. This is particularly noticeable for LDPE-C. For LDPE-E the $\overline{M}_w$ values obtained are just slightly more than 20% higher than the $\overline{M}_w$ value found by light-scattering. Based on this comparison we can only eliminate an exponent of g as high as 3/2.

The table below gives values calculated for the long chain branching frequency, λ, using the two models and different exponents of g.

Method for LCB Frequency		λ ,Branch Points/Unit M.W. , x 10^4	
		LDPE-C	LDPE-E
$g^{1/2}$	(ZS-p)	2.27	1.14
	(ZS-m)	12.2	4.48
$g^{0.7}$	(ZS-p)	0.77	0.51
	(ZS-m)	3.21	1.61
$g^{3.2}$	(ZS-p)	0.15	0.13
	(ZS-m)	0.37	0.29
SCB and LCB Frequency by Infrared		13	14

In this table the total chain branching frequency, as measured by IR, is shown in terms of branch points per unit molecular weight for comparison with the long chain branching frequency. Since the IR method for measuring branching frequency should give a value representing the total short and long branches present in these samples, we do not expect that the true long chain branching frequency will be as high as the short chain branching frequency. The data show in all cases, with the exception of $g^{1/2}$ (ZS-m) case, that reasonable branching frequencies are obtained. On this basis only the $g^{1/2}$ (ZS-m) case can be eliminated from consideration.

The following table shows what happens to the correlation between zero shear viscosity, η_o, and $(\overline{M}g)_w$ when $(\overline{M}g)_w$ is evaluated with these different cases.

$$\ln \eta_o = A + a \cdot \ln(\overline{Mg})_w$$

Method for $(\overline{Mg})_w$		Correlation Coefficient	Slope a
$g^{1/2}$	(ZS-p)	0.99	8.0
	(ZS-m)	0.99	8.0
$g^{0.7}$	(ZS-p)	0.92	5.0
	(ZS-m)	0.92	4.0
$g^{3/2}$	(ZS-p)	0.76	2.6
	(ZS-m)	0.75	2.6

It was shown earlier that our initial choice of $g^{1/2}$ (ZS-p) case gives a slope of about 8.0. The correlation coefficient of 0.99 is extremely good. It can be seen from the above table that the effect of increasing the exponent of g is to progressively decrease the slope. However, if we are to get a slope of 3.5 as predicted by the Bueche equation, we must also take a considerable reduction in the correlation coefficient.

All of the above data taken together suggest that our initial choices of $g^{1/2}$ and the polydisperse model were the best choices, and that the zero-shear viscosity data is best expressed with a higher dependence on $(\overline{Mg})_w$ or $(\overline{S^2})_w$ than predicted by the Bueche expression. The high sensitivity of the slope and correlation coefficient to the exponent chosen for the branching parameter, g, cautions us to be sure that the value of $(\overline{Mg})_w$ which we compute is truly proportional to $(\overline{S^2})_w$. Thus, we need to have a direct measurement of $(\overline{S^2})_w$. Unfortunately, $(\overline{S^2})_w$ is not readily obtained by direct experimental procedures. Light-scattering gives the Z-average value, $(\overline{S^2})_z$ rather than $(\overline{S^2})_w$.

V.   CONCLUSION

From the experimental results of this investigation and others (15, 20-22, 42-43) we can draw several conclusions.

The GPC-IV method of measuring the molecular parameters of LDPE has been successfully demonstrated for fractions and whole polymers. In the GPC-IV method, the best results are obtained by assuming that $g^{1/2} = [\eta]_B/[\eta]_L$ and that each GPC elution is a polydisperse system. The zero shear viscosity of LDPE depends not only on weight-average molecular weight, but also strongly on the long chain branching characteristics. The combined influence of these two molecular parameters on zero shear viscosity can be described through the effect which each has on the weight-average mean square radius of gyration of the polymer coil, $(S^2)_w$. The zero shear viscosities of whole LDPE samples are found to correlate with $(\overline{S^2})_w$ very well. However, the dependence of zero shear viscosity on $(S^2)_w$ is substantially greater for LDPE than for HDPE or other linear polymers. The long chain branching present in LDPE is believed to be responsible for this high sensitivity of zero shear viscosity to a change in $(\overline{S^2})_w$. The apparent flow activation energy at zero shear, E_o, is about 12-13 Kcal/mole for all LDPE samples, whereas E_o is about 6.5 Kcal/mole for HDPE samples. E_o of polyethylene appears to be independent of molecular weight distribution or short chain branching. Long chain branching is believed to be responsible for the high E_o of LDPE. E_o is found to be constant for long chain branching frequencies, λ, above about 1×10^{-4}. There are some indications that E_o increases from about 6 to 12 Kcal/mole with increasing λ below about 1×10^{-4}.

ACKNOWLEDGEMENTS

The authors wish to thank Dr. J. J. McAlpin for his contribution in the initial development of our GPC-IV method for measuring the molecular parameters of low density polyethylene. We also wish to thank Mr. J. O. Brewer for most of the experimental data reported in this paper.

References

1. Bueche, F., J. Chem. Phys., 20, 1959 (1952)
2. Bueche, F., J. Appl. Phys., 26, 738 (1955)
3. Bueche, F., J. Chem. Phys., 25, 599 (1956)
4. Fox, T.G. and S. Loshack, J. Appl. Phys., 26, 1080 (1955)
5. Longworth, R. and H. Morawetz, J. Poly. Sci., 29, 307 (1958)
6. Ninomiya, K., J. D. Ferry and Y. Oyanagi, J. Phys. Chem., 67, 2297 (1963)
7. Schreiber, H. P., E. B. Bagley and D. C. West, Polymer, 4, 355 (1963)
8. Fox, T. G., J. Poly. Sci., C, 9, 35 (1965)
9. Porter, R. S. and J. F. Johnson, Chem. Reviews, 66 (1), 1 (1966)
10. Berry, G. C. and T. G. Fox, Adv. Poly. Sci., 5, 261 (1968)
11. Thomas, D. P., Poly. Eng. Sci., 11 (4), 305 (1971)
12. Bueche, F., J. Chem. Phys., 22, 1570 (1954)
13. Bueche, F., J. Poly. Sci., 41, 549 (1959)
14. Bagley, E. B., J. Appl. Phys., 28, 625 (1957)
15. Tung, L. H., J. Poly. Sci, 46, 409 (1960)

16. Long, V. C., G. C. Berry and L. M. Hobbs, Polymer, 5, 517 (1964)
17. Graessley, W. W., J. Chem. Phys., 43, 2696 (1965)
18. Schreiber, H. P., J. Appl. Poly. Sci., 9, 2101 (1965)
19. Kraus, G. and J. T. Gruver, J. Poly. Sci., A, 3, 105 (1965)
20. Mendelson, R. A., W. A. Bowles and F. L. Finger, J. Poly. Sci., A2, 8, 105 (1970)
21. Wild, L. R. Ranganath and T. Ryle, J. Poly. Sci., A2, 9, 2137 (1971)
22. Drott, E. E. and R. A. Mendelson, J. Poly. Sci., A2, 8, 1361 (1970)
23. Fox, T. G. and V. R. Allen, J. Chem. Phys., 41, 344 (1964)
24. Zimm, B. H. and W. H. Stockmayer, J. Chem. Phys., 17, 1301 (1949)
25. Benoit, H., Z. Grubisic, P. Rempp, D. Decker and J. Zilliox, J. Chem. Phys., 63, 1507 (1966)
26. Grubisic, Z., P. Rempp and H. Benoit, J. Poly. Sci., B, 5, 753 (1967)
27. Krigbaum, W. R. and P. J. Flory, J. Am. Chem. Soc., 75, 1775 (1953):Flory,"Principles of Polymer Chemistry", Cornell, 1953, p 310.
28. Stockmayer, W. H. and M. Fixman, Ann. N.Y. Acad. Sci., 57, 334 (1953)
29. Flory, P. J. and T. G. Fox, J. Am. Chem. Soc., 73, 1904 (1951)
30. Schaefgen, J. R. and P. J. Flory, J. Am. Chem. Soc., 70, 2709 (1948)
31. Thurmond, C. D. and B. H. Zimm, J. Poly. Sci., 8, 477 (1952)
32. Zimm, B. H. and R. W. Kilb, J. Poly. Sci., 37, 19 (1959)
33. Cantow, M., G. Meyerhoff and G. Schulz, Makromol. Chem., 49, 1 (1961)
34. Morton, M., T. Helminiak, S. Godgary and F. Bueche, J. Poly. Sci., 57, 471 (1962)
35. Orofino, T. and F. Wenger, J. Phys. Chem, 67, 566 (1963)
36. Altares, T., Jr., D. Wyman and V. Allen, J. Poly. Sci., A, 2, 4533 (1964)
37. Graessley, W. W. and H. M. Mittelhauser, J. Poly, Sci., A2, 5, 431 (1967)
38. Graessley, W. W., "Detection and Measurement of Branching in Polymers, Characterization of Macromolecular Structure", National Academy of Science, Washington, D.C., 1968, p 371.
39. Chiang, R. J., J. Poly. Sci., 36, 91 (1959)
40. Weissenberg, K., Nature, 159, 310 (1947)
41. Kauzmann, W. and H. Eyring, J. Am. Chem. Soc., 62, 3113 (1940)
42. Ferguson, J., B. Wright and R. N. Haward, J. Appl. Chem., 14, 53 (1964)
43. Boghetich, L. and R. F. Kratz, Trans. Soc. Rheol., 9:1, 255 (1965)

44. Drott, E. E. and R. A. Mendelson, Poly. Preprints, 12(1), 277 (1971)

45. Mendelson, R. A. and E. E. Drott, J. Poly. Sci., B, 6, 795 (1968)

46. Westerman, L. and J. C. Clark, To be published shortly.

THE INFLUENCE OF INTERMOLECULAR HYDROGEN BONDING ON THE OSCILLATORY SHEAR FLOW BEHAVIOR OF THE ETHYLENE-ACRYLIC ACID COPOLYMERS

Timothy Lim and Thomas W. Haas

American Standard, Product, Development & Engineering Laboratory, P. O. Box 2003, New Brunswick, New Jersey, 08903

INTRODUCTION

Polymer melt rheologists have long been concerned with aspects of structure that affect molecular entanglement. However, it has been only recently that there has been interest in secondary forces such as polar and hydrogen bonds.[1-6]

In a recent publication,[4] the results of a high shear rate capillary rheometer study of the flow behavior of copolymers of ethylene and acrylic or methacrylic acid were reported. The hydrogen bonding was found to substantially enhance both the flow activation energy and the level of viscosity as well as the degree of dependence of viscosity on the rate of shear. The proposal was made that the hydrogen bonds act effectively as temporary (quasi-) crosslinks during the short time scales of deformation involved in flow under steady high rates of shear. (>10 sec.$^{-1}$)

The current investigation was undertaken to gather further evidence of the crosslinking action of hydrogen bonds by studying the flow behavior of the ethylene-acrylic acid copolymers in an oscillatory shear deformation.

MATERIALS

The materials studied were ethylene-acrylic acid copolymers of 1.3, 3.1 and 5.3% mole percent acid content. The samples were taken from those characterized previously under high shear in the capillary extrusion rheometer.[4]

   The copolymers are commercially available and are prepared by a high-pressure free radical polymerization. Therefore, they contain both long- and short-chain branching and are
of broad molecular weight distribution. The acrylic acid units
are thought to be distributed at random along the polymer chain. [8]
Infrared measurements indicate that hydrogen bonding persists
in the molten state. [8] Even at 160°C, the highest temperature
employed in this study, some 22%-30% of the carboxyl groups
are associated by hydrogen bonds.

   The properties of the materials investigated in this study are
given in Table I. The polymers are designated by type(BPE for
branched polyethylene and EAA and ethylene-acrylic acid) followed by the acid concentration expressed in mole-% where
appropriate.

TABLE I [4]

Properties of Polymers Investigated

Polymer	Supplier	Acid concen- tration*, Mole-%	Melt Index**	T_β , °C ***	$\frac{CH_3}{100C}$****
BPE	duPont	0	3	---	1. 7
EAA-1. 3%	Union Carbide	1. 3	4. 8	-5	2
EAA-3. 1%	Dow	3. 1	6. 4	12	2
EAA-5. 3%	Union Carbide	5. 3	4. 8	20	1. 5-2. 0

*   Determined by titration
**   ASTM D1238
***   Determined by dynamic loss measurements at 100 cps;
        T_β is associated with the glass transition.
**** Determined by infrared measurements

EXPERIMENTAL

   The oscillatory shear flow experiments were carried out
using the Rheometrics* Mechanical Spectrometer[9] in the eccen-
tric disks otherwise known as orthogonal mode[10, 11] Analysis
of this experiment is present elsewhere.[12, 13] The inphase dy-

*   Rheometrics Inc., Louisville, Kentucky.

namic viscosity, η' , elastic modulus G' and out of phase dynamic viscosity, η'', loss modulus G'' are calculated according to the following equations:

$$G' = 1/A \left(\frac{F'}{\gamma} \right) \tag{1}$$

$$G'' = 1/A \left(\frac{F''}{\gamma} \right) \tag{2}$$

$$\eta' = G''/\omega \tag{3}$$

$$\eta'' = G'/\omega \tag{4}$$

Where: A is the area of the parallel plate
F' is the measurable in-phase force amplitude
F'' is the measurable out of phase force amplitude
ω is the angular frequency
γ is the shear strain

RESULTS AND DISCUSSION

Oscillatory shear flow curves were obtained for the three
acrylic acid copolymers, containing 1.3, 3.1 and 5.3 mole percent acid, at various temperatures. Data taken at 130°C for
these copolymers are shown in Figure 1. The dynamic shear
viscosity, η' , decreases with increasing frequency whereas
the elastic or storage shear modulus, G', increases with increasing frequency which is typical of the viscoelastic response
of polymer melts. Examination of the curves indicates that the
dynamic viscosity is asymptotically approaching a constant
value which defines the zero shear viscosity, η_o;

$$\eta_o = \lim_{\omega \to 0} \eta'$$

where ω is the frequency.

The zero shear viscosity of the copolymers was estimated
by plotting the reciprocal of the dynamic viscosity, $1/\eta'$, versus the shear loss modulus, G''. As G'' approaches zero, $1/\eta'$,
approaches $1/\eta_o$. This is an adaptation of Ferry's method used
for steady flow experiments.[14]

The zero shear viscosity affords an advantage because it is
determined uniquely by the temperature, i.e., it is independent
of shear rate or frequency. The temperature dependence of the
zero shear viscosity can be described as flow activation energy,
E_o. The zero shear flow activation energy E_o is defined by the

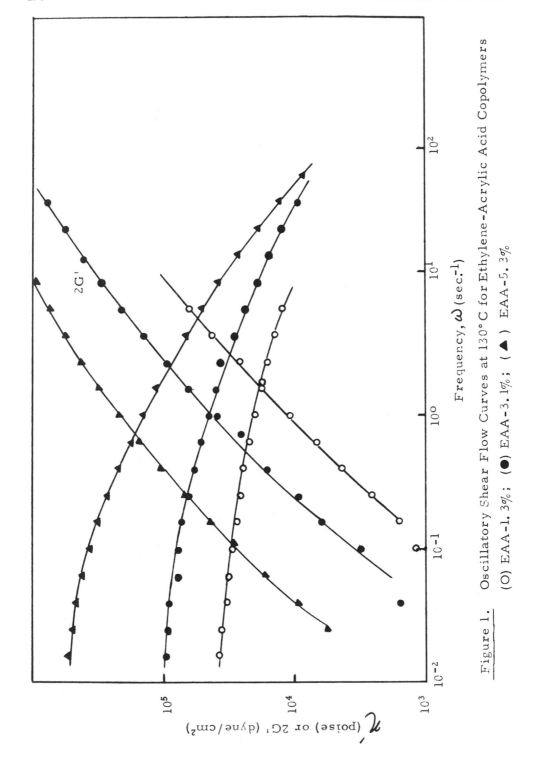

Figure 1.   Oscillatory Shear Flow Curves at 130°C for Ethylene-Acrylic Acid Copolymers

(O) EAA-1.3%;   (●) EAA-3.1%;   (▲) EAA-5.3%

Arrhenius equation:

$$\eta_o(T) = A \cdot \exp\left[E_o/RT\right]$$

where A is a constant, R is the gas constant, and T is the absolute temperature.

The flow activation energy of the zero shear viscosity of
each of the acrylic acid copolymers along with the results obtained by similarly treating oscillatory shear flow data for a
sample of branched polyethylene (BPE) are shown in Table II.
Also given are the steady flow activation energies in the capillary rheometer under a shear stress of 1×10^6 dynes/cm^2. The
incorporation of hydrogen bonds by means of increasing acrylic
acid content in the copolymers results in an increase in the flow
activation energy over that of branched polyethylene both at the
high shear rates employed in the capillary rheometer and at the
low shear rates of the zero shear viscosity. The contribution
of the hydrogen bonds to the flow activation energy is therefore
independent of the rate of shear.

TABLE II

Comparison of the Activation Energy Data Obtained from
both Steady Shear and Oscillatory Shear Flow

Polymer	E_o (Oscillatory Flow Acti- vation Energy at Zero Shear)	E_γ (Steady Shear Activa- tion Energy Measured at 1×10^6 dyne/cm^2)
BPE	14.1 Kcale/g-mole	10.4 Kcale/g-mole
EAA-1.3%	16.1 "	12.7 "
EAA-3.1%	16.2 "	14.0 "
EAA-5.3%	18.1 "	15.7 "
EAA-5.3% 85% ester	10.4 "	8.5 "

The dynamic flow results were obtained by applying a small
oscillatory deformation to the polymer melt well within the linear viscoelastic range whereas the previous results were obtained
in a capillary rheometer in which the polymer undergoes a very
large, steady deformation. The fact that the flow activation

energy depends upon the acrylic acid content in a similar manner for both types of experiments implies that the effects of hydrogen bonding are independent of the magnitude of the deformation.

With further examination of the dynamic flow curves in Figure 1, there appears to be a regular increase in both the dynamic viscosity and the shear modulus with increasing acrylic acid content. However, although the melt indices of the three acrylic acid copolymers are comparable, the foregoing observation may be fortuitous owing to differences in molecular weight, molecular weight distribution and long-chain branching.

In order to demonstrate the effect of hydrogen bonds on the level of dynamic viscosity and the elastic modulus, experiments must be performed on identical molecular structures having only different degrees of intermolecular association present. This was accomplished in the capillary flow study[4] by esterification of one of the acrylic acid copolymers, EAA-5.3%, by reaction of the carboxyl groups with diazomethane which destroyed 85% of the hydrogen bonds. The capillary flow curves revealed that in addition to an increase in the activation energy for viscous flow, hydrogen bonding leads to an enhancement of the level of viscosity and the degree of dependence of the viscosity on the rate of shear as well. This was taken as evidence that hydrogen bonds act as temporary (quasi-) crosslinks at the short time scales of deformation involved in flow under high rate of shear.[4]

The oscillatory flow curves for the esterified copolymer (EAA-5.3%, 85% ester) at 130 and 160°C are shown in Figure 2. Also shown are the curves for the copolymer (EAA-5.3%). In the absence of any effect of hydrogen bonding, these dynamic flow curves would be expected to be virtually identical. However, both the dynamic viscosity and the elastic modulus are higher for the EAA-5.3% sample than for the esterified sample. Furthermore, the dynamic viscosity of the EAA-5.3% also exhibits a greater degree of frequency dependence in agreement with the result obtained in capillary flow.

If one were dealing with polymers differing in only molecular weight, a result such as that above would be attributed to the presence of varying degrees of physical crosslinks, i.e., molecular entanglements between the individual polymer chains, and interpreted in terms of the theories of the molecular weight (chain length) dependence of viscosity[18] and rubber elasticity.[19] In this case, the fact that the EAA-5.3% sample exhibits much higher viscosity, shear modulus and greater degree of shear dependence than the esterified sample having equivalent molecular weight, provides strong evidence for the temporary crosslinking effect of the hydrogen bonds. The flow activation energy for the esterified

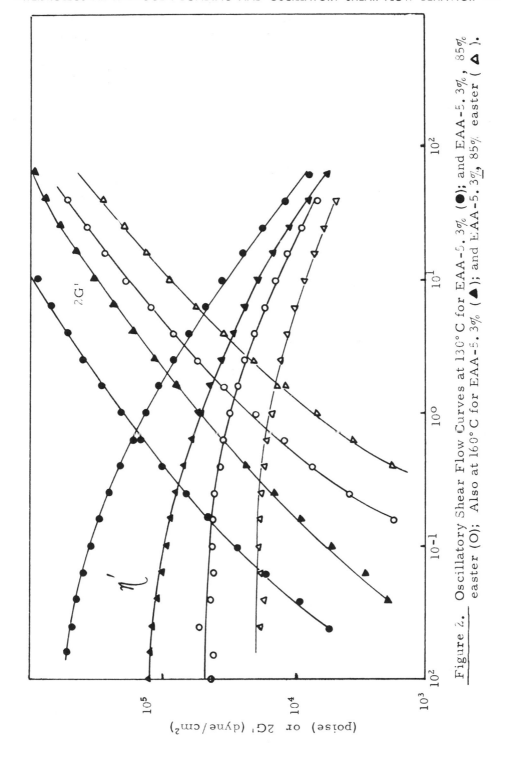

Figure 2. Oscillatory Shear Flow Curves at 130°C for EAA-5.3% (●); and EAA-5.3%, 85% easter (O); and EAA-5.3%, 85% easter (▲); and EAA-5.3%, 85% easter (△). Also at 160°C for EAA-5.3% (▲); and EAA-5.3%, 85% easter (△).

sample, FAA-5.3%, 85% ester and the control simple, EAA-
5.3% are shown in Table II. The control has an activation energy
nearly twice that of the esterified sample. This is also true in
the case of the capillary experiment. This finding gives further
evidence of the influence of hydrogen bonding on the flow behavior
of polymer melts.

The flow activation energy of the esterified sample, however,
is not very close to that of the branched polyethylene (BPE). If
other molecular structural parameters are equal, the esterified
sample should have an activation energy approximately equal to
that of the branched polyethylene. One possible explanation is that
the branched polyethylene may contain more long-chain branching
than that of the esterified sample. Experimental evidence[15-17]
indicates that long-chain rather than short-chain branching is responsible for the increase in activation energy.

As mentioned above, the contribution of the hydrogen bonds to
the flow activation energy is independent of the rate of shear. On
the other hand, the effect of long-chain branching on the flow activation energy is quite dependent on the rate of shear. The evidence
is seen in the stress dependence of the activation energy of
branched polyethylene,[4] for as stress is increased, a corresponding drop in E_a is observed. This interpretation provides the explanation for the fact that the difference in the flow activation
energies between the esterified sample and the branched polyethylene is found to be larger in the low shear flow than in the
high shear of the capillary. (See Table II)

CONCLUSIONS

The increase in the flow activation energy of the acrylic
acid copolymers over that of branched polyethylene both at the
high shear rates employed in the capillary rheometer[4] and the
low shear rates at the zero shear viscosity indicates that the
contribution of the hydrogen bonds to the flow activation energy
is independent of shear rate. Increased hydrogen bonding leads
to an enhancement of the level of the dynamic viscosity and the
degree of dependence of the dynamic viscosity on frequency.
This is in agreement with the previous results obtained for the
steady viscosity in capillary flow. Furthermore, the elastic
shear modulus increases with increased acrylic acid content.
The experimental data indicate that intermolecular hydrogen bonds
act effectively as temporary (quasi-) crosslinks during the time
scales of deformation both at high and low rates of shear.

REFERENCES

1. R. Longworth and H. Morawitz,
    J. Poly. Sci. , 29. 307 (1958)

2. W. E. Fitzgerald and L. E. Neilsen,
    Proc. Roy Sci, Ser. A, 282, 137 (1962).

3. R. J. Boyce, W. H. Bauer, and E. A. Collins,
    Trans. Sci. Rheol. , 10 (2). 545 (1966).

4. L. L. Blyler, Jr. and T. W. Haas,
    J. Appl. Poly. Sci. , 13, 2721 (1969).

5. E. A. Collins, T. Mass, and W. H. Bauer,
    Rubber Chemistry and Technology, 43, 1109 (1970).

6. K. Sakamoto, W. J. MacKnight, and R. S. Porter,
    J. Poly. Sci. A-2, 8, 277 (1970).

7. R. Longworth, private communication

8. E. P. Otocka and T. K. Kwei,
    ACS Polymer Preprints, 9, 583 (1968).

9. C. Macosko and J. M. Starita, SPE Journal,
    27, 38 (1971).

10. B. Maxwell and R. P. Chartoff,
    Trans. Soc. Rheology, 9, 41 (1965).

11. B. Maxwell, Poly. Eng. and Sci. ,
    8, 257 (1968).

12. L. L. Blyler, Jr. and S. J. Kurtz, J. Appl.
    Poly. Sci. , 11, 127 (1967).

13. R. B. Bird and E. K. Harris, Jr. AIChE J ,
    14, 758 (1968).

14. L. L. Blyler, Jr. , Trans. of the Soc. of Rheology,
    13, 39 (1969).

15. L. H. Tung, J. Polymer Sci. , 46, 409 (1960).

16. R. Sabia, J. Appl. Polymer Sci. , 8, 1651 (1964).

17. R. A. Mendelson, Trans. Soc. Rheol. , 9 (1), 53 (1965).

18. J. D. Ferry, "Viscoelastic Properties of Polymers,"
     2nd Fdition, Wiley, 1970.

19. L. R. G. Treloar, "The Physics of Rubber Flasticity,"
     2nd Fdition, Oxford, 1958.

SURFACE MORPHOLOGY AND DEFORMATION BEHAVIOR OF POLYPROPYLENE SINGLE CRYSTALS

D. R. Morrow, R.H. Jackson* and J. A. Sauer

Department of Mechanics and Materials Science

Rutgers, The State University of New Jersey

ABSTRACT

As part of an investigation into the nature and properties of crystalline polymers, the morphology and deformation behavior of polypropylene (PP) single crystals have been examined. The study includes single crystals of the whole polymer as well as single crystals grown from each of three PP fractions ($\overline{M}n$ = 740, 1260 and 1880). The structure of the crystal surfaces and the mechanical response of each of the single crystals to a tensile strain have been examined. Results have been obtained which indicate how the morphology of the crystal surface may be altered through the use of specific thermal and/or chemical treatments. A correlation has been found to exist between the surface morphology and deformation behavior of the PP single crystals. Two possible mechanisms are proposed to explain this relation between these single crystal properties. It is shown that the combined use of metal decoration procedures and a quantitative analysis of the deformation behavior of PP single crystals provides a powerful technique for studying the properties of polymer crystal surfaces.

INTRODUCTION

The presence of molecular chain-folding in both solution and bulk-crystallized polymeric materials is well established as a common feature of crystalline polymers. The most elementary unit of a crystalline polymer is the single crystal. Therefore, in order to improve our understanding of polymer crystals, and more

* Celanese Fibers Co., Narrows, Virginia

specifically polypropylene single crystals, it is important to gain insight into (a) the nature of chain-folding in polymer crystals, (b) the morphology of the fold surface of polymer crystals, and (c) the influence of chain folds on the physical properties of polymer crystals. For this purpose whole polymer crystals (WPC) of polypropylene (PP) were grown by isothermal crystallization from dilute solution of isotactic PP in α-chloronaphthalene (α-CN) as previously reported (1). Crystals were also grown by solvent evaporation from a dilute paraxylene solution using PP fractions of molecular weight ($\overline{M}n$) 740, 1260, and 1880 (2). Composite crystals were grown in which the WPC were used as the nucleation sites for the epitaxial growth of the 740 crystals.

Due to the small size of the PP single crystals, conventional deformation techniques cannot be employed. One technique which has proven to be quite successful is to deposit or grow the single crystals on a substrate which is then subjected to an applied, tensile strain (3). The strain is then transmitted from the substrate to the crystals by the interaction of their surfaces. It is reasonable to expect the structure of the crystal surfaces to influence the nature of the surface interaction and, hence, the deformation response. Therefore, it is desirable to study the nature of the surface of the PP single crystals as well as their deformation behavior.

EXPERIMENTAL MATERIALS AND TEST PROCEDURE

The whole polymer crystals of PP were obtained by first washing the polypropylene powder (Avisun) with boiling n-heptane to remove any atactic polymer present. A 0.004% (by weight) solution of the isotactic polymer in α-CN was prepared. This solution was first heated to 190°C to dissolve the polymer and the solution was then cooled rapidly to 112°C. It was maintained at this temperature for a period of three to five days until crystallization was completed.

Crystals of the low molecular weight fractions were prepared by dissolving a small amount of the fraction in para-xylene to form a 0.1% (by weight) solution. A drop of this "stock" solution was placed on the desired substrate and the polymer was crystallized by solvent evaporation. The crystallization conditions were those reported previously for these fractions (2).

To obtain a composite crystal, the WPC were first deposited on the desired substrate. A small amount of the stock solution of the 740 m.w. fraction was then deposited on the substrate and allowed to crystallize by solvent evaporation at ambient temperature and pressure. The result was the epitaxial growth of the 740 crystal from the end face (010) of the WPC in the manner already reported (4).

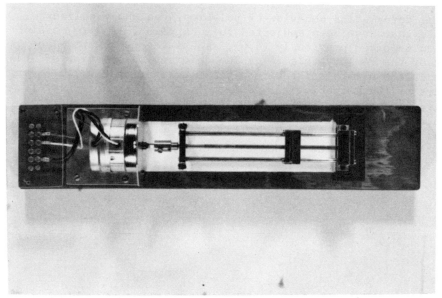

Figure 1   Photograph of motor-driven mechanical stretching device

To determine the modes of deformation of the single crystals, the desired crystals were deposited or grown on a Mylar substrate. This substrate was mounted in a motor-driven mechanical stretcher (Fig. 1). All deformation tests were conducted at the same strain rate -- approximately 1% strain per minute. The sample preparation technique employed was essentially that previously reported by Geil (5).

The surface texture study involved a comparative examination of the surface textures of the 740 crystals relative to the whole polymer crystals. To determine what parameters affect the surface texture of the WPC, various treatments were applied to the WPC prior to the epitaxial growth of the 740 crystals. The 740 composite crystals were shadowed with Pd-Au to enhance the surface texture. All of the samples were shadowed at an angle to the substrate of $\tan^{-1}(1/3)$.

The negatives of the micrographs obtained in the surface texture study were scanned with a microphotometer to determine the relative granularity of the crystal surfaces. The negatives each have about the same magnification (43,000X), and were scanned at the same scan rate and with the same slit width. The ratio of the number of grains per unit length for the 740 and whole polymer crystals is reported as the specific granularity.

RESULTS

The deformation behavior of polypropylene whole polymer
crystals that have been subjected to an applied, tensile strain
has been reported previously (4,6), as has the deformation behavior of the 740 crystals (4). The deformation behavior of the
1260 and 1880 crystals is quite similar to that of the 740 crystals
in terms of the morphological changes in the strained crystal.
That is, the extended-chain crystals of the 740, 1260, and 1880
molecular weight fractions exhibit periodic cleavages of the
crystal normal to the draw direction when they are strained
(Fig. 2). However, the folded-chain crystals of the whole polymer

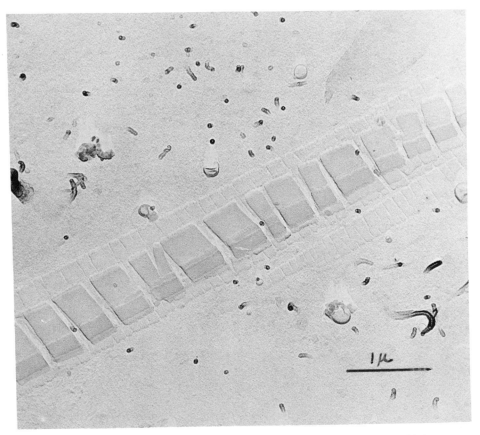

Figure 2   Electron photomicrograph of 1880 crystals on Mylar,
           strained 25%. Shadowed with C-Pt.

exhibit periodic (micro)craze-cracks with fibrils extending across
the (micro)craze-cracks when they are strained. Of importance in
these results is the relation between the induced and applied
strain for the various crystals. The applied strain is defined in
terms of the amount the deformable substrate is strained. The
induced strain is defined as the strain measured directly for the
deformed crystals and is a function of the ratio of crack width and
undeformed crystal segment length. In the case of the WPC, a
correction is made in the induced strain that is measured in an
effort to account for the material which is drawn out of the undeformed crystal segments to form the fibrils. This correction term
is discussed in the Appendix.

As stated above, a qualitative analysis of the deformation
behavior of the whole polymer crystals and 740 crystals subjected
to a tensile strain has already been reported (4). Figure 3 shows

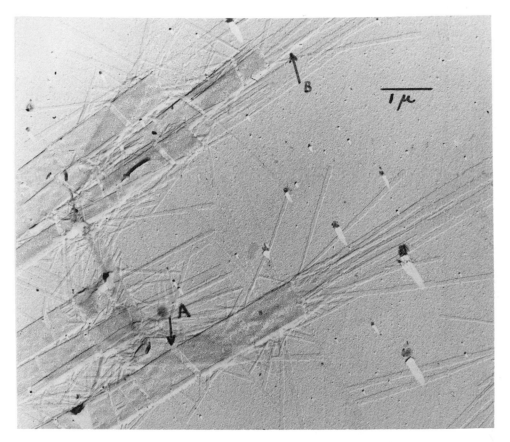

Figure 3 Electron photomicrograph of 740 composite crystals on
         Mylar, strained 10%. Shadowed with Pd-Au.

a representative composite crystal which has been subjected to a strain of 10%. Some (micro) craze-cracks can be seen in the whole polymer crystal portion of the composite crystal with fibrils extending across these (micro) craze-cracks (Figure 3, Arrow "A"). Fractures in the 740 crystal portion of these composite crystals can also be seen (Arrow "B").

From a quantitative analysis of the deformation behavior of the 740, 1260, 1880, and whole polymer crystals, the graph in Figure 4 can be drawn. This figure shows that there is a 1:1

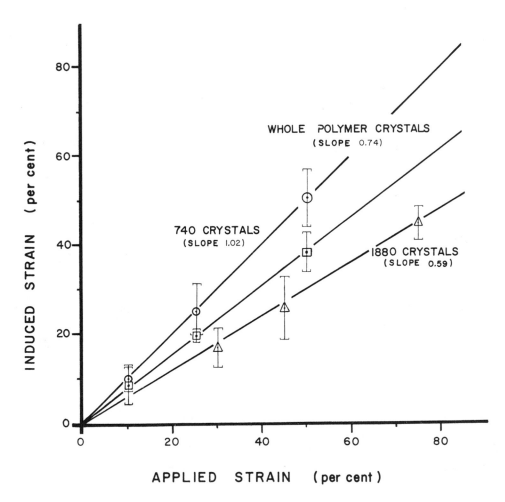

Figure 4   Induced strain vs. applied strain for the 740, 1880, and whole polymer crystals

correspondence between the induced and applied strain for the 740
crystals. Although it is not recorded in this figure, the 1260
crystals also have a 1:1 correspondence between the induced
and applied strain. The WPC have an induced strain which is 74%
of the applied strain, and the 1880 crystals have an induced strain
which is 59% of the applied strain. Some WPC were annealed prior
to being strained. These crystals were annealed from 0.5 to 1
hour at 150 – 160°C under vacuum, cooled slowly to ambient temperature, and then drawn in the mechanical stretcher. Figure
5 shows that the annealed whole polymer crystals exhibit a 1:1

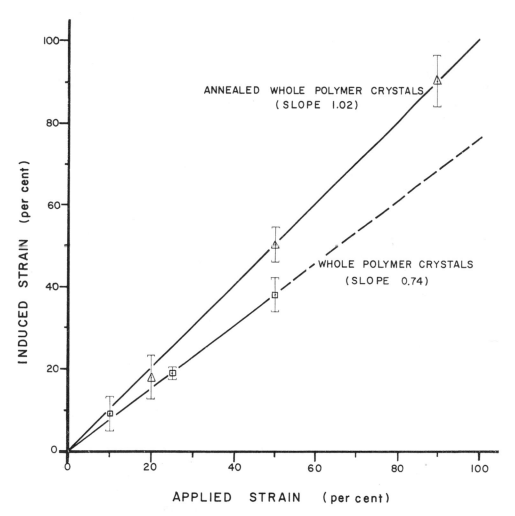

Figure 5   Induced strain vs. applied strain for the whole polymer
            crystals and the annealed whole polymer crystals.

correspondence between the induced strain and applied strain just as the 740 and 1260 crystals do.

The error bars in Figure 4 and 5 represent the standard deviation from the mean. The slope of the lines is based on a least squares fit of the data. A statistical analysis of the data indicates that, with a confidence limit of 95%, there is a significant difference between the slopes of the lines for the 740, 1880, and whole polymer crystals. A similar analysis indicates that there is no significant difference between the slopes for the 740, 1260, and annealed whole polymer crystals.

Figure 6 shows a representative composite crystal (C-Pt

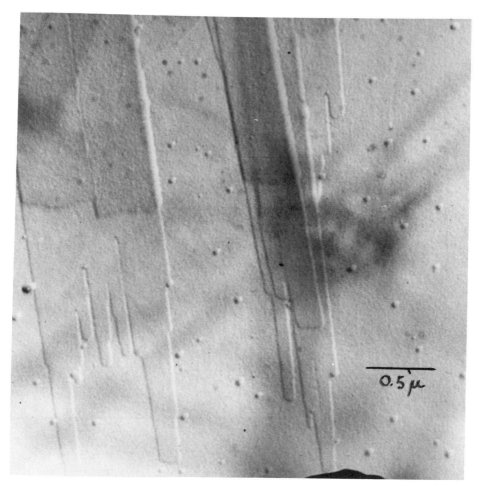

Figure 6 Electron photomicrograph of 740 composite crystals shadowed with C-Pt.

Shadowed) for which a folded-chain, whole polymer crystal was used as a nucleation site for the epitaxial growth of an extended-chain crystal of the 740 fraction. It can be seen that the whole polymer crystal has a surface which displays more textural detail than the surface of the 740 crystal.

The 740 composite crystals in Figure 7 have been shadowed with Pd-Au, and this particular shadowing material is seen to en-hance the surface texture. Here it can be seen that the texture

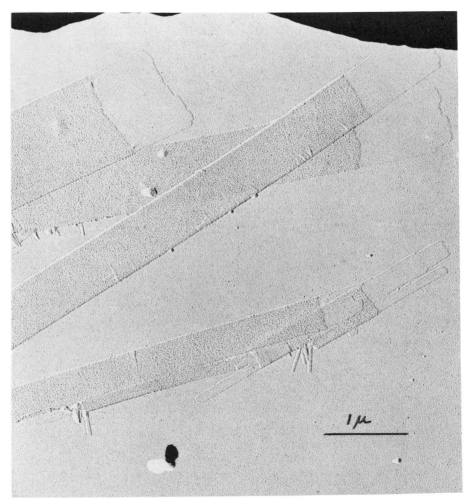

Figure 7 Electron photomicrograph of 740 composite crystals shadowed with Pd-Au.

of the 740 crystal portion is essentially indistinguishable from that of the carbon substrate background. The WPC surface possesses more textural detail than the 740 crystal surface or background.

The whole polymer crystals shown in Figure 8 were annealed at

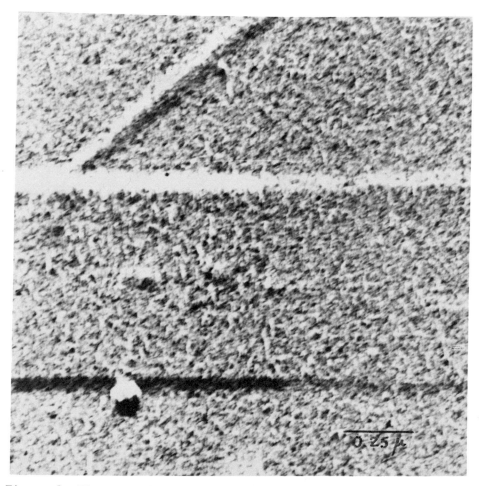

Figure 8   Electron photomicrograph of whole polymer crystals deposited on a carbon substrate, annealed at 147°C for 1 hour, cooled slowly, and then 740 crystals were grown onto the annealed whole polymer crystals. Shadowed with Pd–Au.

147°C in a vacuum for 1 hour prior to the epitaxial growth of the 740 crystals. Based on previous annealing studies, this amounts to a relatively light heat treatment (7). Both portions of the composite crystal appear to have surfaces with the same amount of textural detail.

Figure 9 shows whole polymer crystals that were washed with

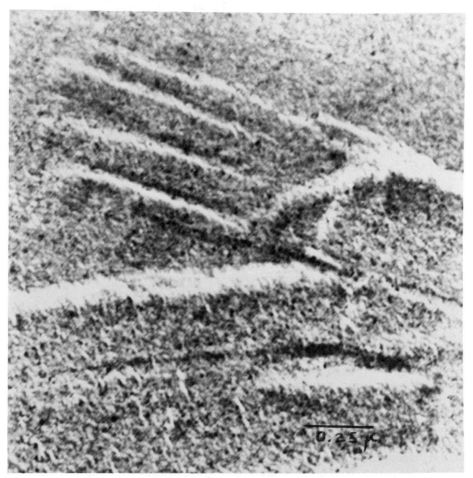

Figure 9 Electron photomicrograph of whole polymer crystals de-
          posited on a carbon substrate, washed with α-CN at 110°C
          in air, cooled rapidly. 740 crystals were grown onto the
          washed whole polymer crystals. Shadowed with Pd-Au.

hot solvent, α-CN, at 110°C in air followed by rapid cooling. The whole polymer crystals appear to have a greater amount of textural detail than the 740 crystals.

When the whole polymer crystals were washed with α-CN at 110°C for 1/2 hour and cooled slowly, there was no apparent difference in surface texture for the two portions of the composite crystal (Figure 10). This is the same result found for the whole polymer crystals that were given a light heat treatment (Figure 8).

The composite crystals in Figure 11 were strained 25% and shadowed with Pd-Au normal to the draw direction. The straining does not appear to have altered the relative surface textures from what is observed for the untreated crystals.

The negatives of Figures 7, 8, 9, 10, and 11 were scanned with the microphotometer to determine the relative granularity of the crystal surfaces. The ratio of the number of grains per unit length for the 740 crystals to grains per unit length for WPC is defined as the specific granularity. The results of these tests (Table 1) substantiate the qualitative observations of the relative surface textures already stated.

DISCUSSION

Correlating the results of the study of the deformation behavior and surface texture of the polypropylene single crystals, it is possible to propose two mechanisms which relate these properties. These may be designated as the mobile-surface mechanism and the contact-area mechanism.

Before discussing possible deformation mechanisms, it is important to consider the relative melt properties of the low molecular weight fractions as determined by the DSC. Since the thermal treatment prior to and during the DSC scan is similar for each fraction, the width of the melt peak is associated mainly with the molecular weight distribution within each fraction. From Table II it can be seen that the melt peak for the 1880 fraction is 3 1/2 times wider than the melt peak for the 740 fraction. Since the 740 fraction is relatively sharp, the chains will all be of approximately the same molecular weight (chain length). The extended-chain crystals grown from this fraction will have relatively few chains longer or shorter than the theoretical length of 38 Å. Thus at the surface of the crystal, the chain ends in the 740 crystal will have relatively little mobility as a result of their uniform length, and the surface would also be expected to possess relatively little textural detail. However, the width of the melt peak of the 1880 fraction indicates that there are a substantial number of chains of greater and lesser molecular weight than 1880 in the fraction. This means that there will be a significant number of chains which

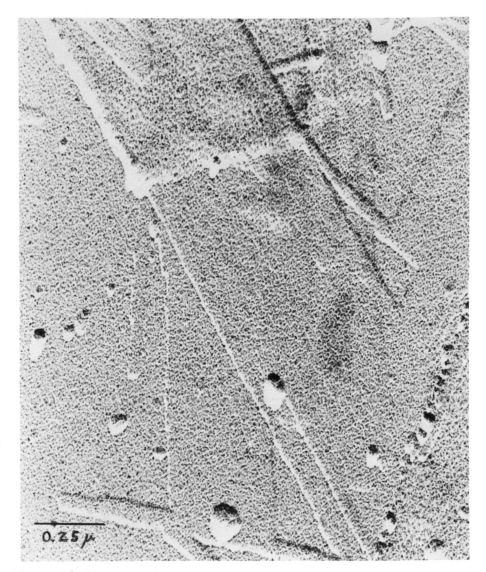

Figure 10 Electron photomicrograph of whole polymer crystals deposited on a carbon substrate, washed with α-CN at 110°C
for 1/2 hour, cooled slowly. 740 crystals were grown
onto the washed whole polymer crystals. Shadowed with
Pd-Au.

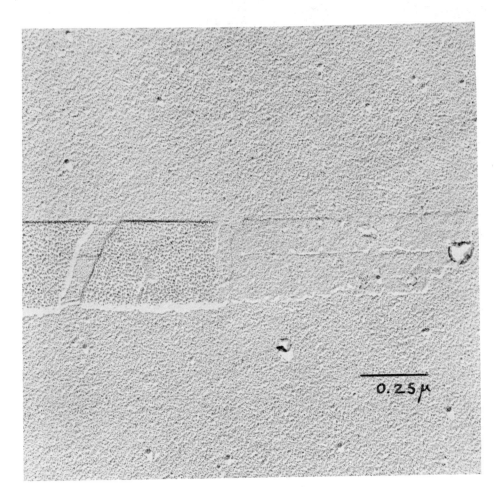

Figure 11 Electron photomicrograph of 740 composite crystals on Mylar, strained 25%. Shadowed with Pd–Au.

RELATIVE SPECIFIC GRANULARITY OF 740 COMPOSITE CRYSTALS

FIGURE NUMBER	TREATMENT GIVEN TO WHOLE POLYMER CRYSTAL OF 740 COMPOSITE CRYSTAL	MAGNIFICATION OF NEGATIVE	SPECIFIC GRANULARITY	AVERAGE SPECIFIC GRANULARITY
107	VIRGIN CRYSTALS	43,000	run 1 1.45 run 2 1.46 run 3 1.46 run 4 1.43	1.45
108	VIRGIN CRYSTALS – REPRODUCIBILITY SAMPLE	43,000	run 1 1.42	1.42
110	ANNEALED AT 147°C FOR 1 HOUR UNDER VACUUM	43,000	run 1 1.00 run 2 0.89 run 3 1.08	0.99
111	WASHED WITH α-CN AT 110°C IN AIR, COOLED RAPIDLY	43,000	run 1 1.60 run 2 1.60 run 3 1.70	1.63
112	WASHED WITH α-CN AT 110°C FOR 1/2 HOUR, COOLED SLOWLY TO AMBIENT TEMPERATURE	43,000	run 1 1.09 run 2 1.03 run 3 1.04	1.05
113	740 COMPOSITE CRYSTALS ON MYLAR, STRAINED 25%	41,500	run 1 1.38 run 2 1.40 run 3 1.45	1.41

TABLE I

D. R. MORROW, R. H. JACKSON, AND J. A. SAUER

TABLE II

DSC MELT PEAK CHARACTERISTICS (°C)[*]

SAMPLE	START	APEX	END	PEAK WIDTH
740	90	92	94	4
1260	101	105	109	8
1880	126	133	140	14
ISOTACTIC POLYPROPYLENE	133	159	167	34

[*]UNPUBLISHED DATA OF DR. B. A. NEWMAN FOR THE POLYPROPYLENE FRACTIONS ($\overline{M}n$= 740, 1260, AND 1880)

which are longer and shorter than the theoretical length of 85Å for this fraction. Thus, the chain ends at the surface of an 1880 crystal will be more mobile due to the variation in chain lengths throughout the crystal. This would impart a more mobile surface to the crystal.

As mentioned, the variation in molecular weight (chain length) distribution present in each fraction is reflected also in the relative textural detail of the crystal surface. That is, assuming that molecular weight segregation is unlikely in this case, the wider the distribution of molecular weights or chain lengths present in any fraction, the less likely it is that the crystal grown using that fraction will have a highly ordered surface free of textural detail. Therefore, the surface of an 1880 crystal would be expected to possess more textural detail than the surface of a 740 crystal. This statement appears to be supported by the results observed for 1880 crystals.

Consider, first, the mobil surface mechanism. Based on the
relative distribution of chain lengths for the respective fractions, as indicated by their DSC melt peaks, it can be concluded
that the 740 crystals possess a surface with relatively low
mobility while the 1880 crystals have a more mobile surface. The
mobile surface mechanism assumes that there is no slippage during
deformation, that is there is no movement of the lower crystal
surface with respect to the substrate surface. The mobile surface
allows the crystal to absorb some of the applied strain so that not
all of the strain is transmitted to that portion of the crystal
above the mobile surface. The mobility of the crystal surface
appears to be related to the amount of textural detail present.
Annealing the WPC initiates a limited refolding of the chains in
the crystal. This appears to reduce the amount of textural
detail of the crystal surface so that the mechanical response of
the annealed whole polymer crystals is the same as that of the
740 crystals in terms of induced strain. Washing the WPC with
hot solvent and then quenching does not allow for a reordering of
the surface layer. However, washing the crystals with hot solvent,
followed by slow cooling gives the surface sufficient mobility
for it to reorder as the crystals are cooled slowly. This will
have the same effect on the crystal surfaces as a light annealing.
Straining the crystal does not extend the order of the crystal
surface so that deformation has no effect on the surface texture
of the whole polymer crystals, and thus has no effect on the
mobility of the crystal surface.

Consider the second mechanism which is called the contact area
mechanism. Those crystals with the least amount of surface
textural detail will have the most surface area in contact with
the substrate material. The frictional forces acting at the
molecular level, which hold a crystal in contact with the surface
of a substrate are proportional to the contact area. Hence
the 740 crystals will have a greater contact area than the 1880
crystals and, therefore, will be less likely to slip. Annealing
the WPC, or washing them with hot solvent followed by slow cooling,
reduced the amount of textural detail of the surface so that the
contact area is increased. Therefore, the annealed WPC should slip
less than the virgin WPC when strained, which is what was observed.

It is likely that at low strain levels (<5%) the mobile surface
mechanism will dominate because the mobile surface allows the
crystal to absorb a portion of the applied strain, while breaking
a minimum number of "bonds" between the crystal surface and substrate surface. The amount of strain that such a mobile surface
can absorb will be a function of temperature and depth of the mobile
surface layer. It should be emphasized that this is a low strain
mechanism and is most likely an elastic response to the applied
strain.

When the limit of extensibility of the mobile surface is reached, it is expected that the contact area mechanism will influence the deformation behavior and determine the amount of slip that will occur. The contact area mechanism is thus a plastic or visco-elastic response to the applied strain.

In conclusion, the results of this investigation indicate that when PP single crystals adhering to a substrate are subjected to a tensile strain, their deformation behavior is strongly dependent upon the nature of the crystal surface. The textural detail of the WPC surface can be reduced by a light annealing or washing with hot solvent followed by slow cooling. Of further importance is the fact that an effective technique has been devised for studying the properties of polymer surfaces. That is, a correlation has been established between the surface morphology, as revealed by microscopy, and the deformation behavior for PP single crystals. With a thorough knowledge of the deformation behavior of a polymer single crystal it is possible to determine the morphology of the crystal surface when direct measurement or observation is not possible. This places additional importance on deformation studies of polymer crystals since these results can be used as a tool for studying crystal surfaces.

APPENDIX

The response of the polypropylene single crystals to an applied strain is defined as the induced strain. For the single crystals grown from one of the low molecular weight fractions of polypropylene studied, the induced strain is the ratio of the crack width (Δl_i) to undeformed crystal segment length (l_{oi}). Referring to Figure 12 this may be expressed as

$$E_i = (\Delta l_i / l_{oi}) \times 100$$

where

E_i = induced strain (per cent)

l_{oi} = length of an undeformed crystal segment for a drawn crystal.

Δl_i = average distance between undeformed crystal segments on either side of a given undeformed crystal segment.

For the whole polymer crystals the induced strain that is measured is defined as:

$$E_{i_m} = (\Delta f / l_o'') \times 100 \qquad (1)$$

where, referring to Figure 13,

Δf = average of the length of the fibrils on either side of an undeformed crystal segment

l_o'' = length of the associated undeformed crystal segment.

Since this definition of induced strain is based on the length of the undeformed crystal segment after straining, it does not take into account the change in length of this segment due to the formation of fibrils. Therefore, this definition of induced strain must be corrected so that it is based on the original length of the undeformed crystal segment.

Assuming that the fibrils are composed of fully extended molecules, in forming the fibrils,

$$nd = \Delta f \qquad (2)$$

where

n = integer number of chains which have completely unfolded to form fibrils of length Δf

d = thickness of the whole polymer crystal,

and

$$ma = l_o' - l_o'' \qquad (3)$$

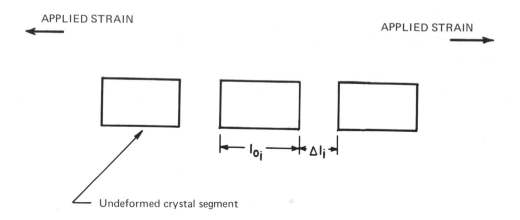

Figure 12   Schematic representation of deformation of extended
             chain polymer single crystals.

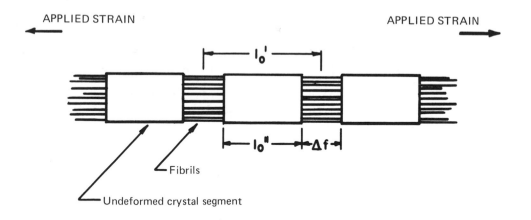

Figure 13   Schematic representation of deformation of lath-
             shaped polypropylene single crystals.

where

> m = integer number of folds lost due to the formation of fibrils (equal to n of Equation 1 for the case of fully extended chains in the fibrils).

> a = unit cell parameter which is equal to twice the chain separation distance to account for the material being removed from both ends of the undeformed crystal segment.

> $1'_o$ = length of the undeformed crystal segment before straining.

> $1_o''$ = length of the undeformed crystal segment after straining.

Hence, solving for n and m, and equating the results, the length of the lamellae consumed to form the fibrils is:

$$1'_o - 1_o'' = (a/d) \Delta f. \qquad (4)$$

The induced strain based on the original length of the undeformed crystal segment, may be expressed as

$$E_{i_c} = \frac{(1_o'' + \Delta f) - 1'_o}{1'_o} \times 100 \qquad (5)$$

Substituting (4) into (5)

$$E_{i_c} = \frac{\Delta f}{1'_o} (1 - a/d) \times 100. \qquad (6)$$

The error in defining the induced strain based on the undeformed length of the strained crystal is reflected in the ratio of induced strain (measured) (E_{i_m}) to the induced strain (corrected) (E_{i_c}). Dividing (6) by (1), substituting (4) into the result and rearranging, the induced strain (corrected) may be calculated as,

$$E_{i_c} = \frac{(1 - a/d) \times 100}{100 + (a/d) E_{i_m}} \times E_{i_m}. \qquad (7)$$

For the virgin whole polymer crystals, the average thickness is

$$d = 125 \text{ Å}$$

and for the annealed whole polymer crystals the average thickness is

$$d = 200 \text{ Å}$$

For both types of whole polymer crystals

$$a = 6.65 \text{ Å} (4).$$

With these values substituted into Equation (7), it is found
that an error of about 10% is incurred for the virgin whole polymer
crystals and an error of about 5% for the annealed crystals if the
induced strain is based on the length of the undeformed crystal
segment after deformation rather than before (for an applied
strain of 100%). It is this corrected value of induced strain
which is used when referring to the induced strain in the text.

This is the first approximation to a correction for the
amount of material drawn out of the undeformed crystal segments
to form fibrils for the strained whole polymer crystals. If it is
assumed that the fibrils are composed of folded-chain segments
rather than extended chains, then even more of the undeformed
crystal segment will be consumed to form the fibrils. As a result,
the size of the correction term will be increased. Therefore,
the correction term presented represents a minimum correction
that must be considered.

ACKNOWLEDGEMENTS
This research was made possible through the support of the
Textile Research Institute, Princeton, New Jersey and the Na-
tional Science Foundation.

REFERENCES
1. D. R. Morrow, J. A. Sauer, and A. E. Woodward,
    J. Polymer Sci., Part C, 16, 3401-11 (1968).

2. D. R. Morrow and B. A. Newman, J. Applied Phys., 39,
    11, 4944-50 (1968).

3. P. H. Geil, J. Polymer Sci., A2, 3813, 3835, 3857 (1964).

4. R. H. Jackson, D. R. Morrow, and J. A. Sauer, A.C.S.
    Preprints, Sept., 1969.

5. P. H. Geil, Polymer Single Crystals, Interscience, New
    York, 1963.

6. P. Cerra, D. R. Morrow, and J. A. Sauer, J. Macromol. Sci., -
    Phys., B3(1), 33-51 (March 1969).

7. G. C. Richardson, "Some Thermal Effects on the Morphology
    of Polypropylene & Poly 4-Methylpentene-1 Single Crystals,"
    M. S. Thesis Rutgers University 1966.

STRUCTURE AND PROPERTIES OF CELLULOSE ACETATE MEMBRANES

James Dickson

Airco Central Research Laboratory

Murray Hill, New Jersey 07974

INTRODUCTION

Reverse osmosis (RO), a technique used to remove unwanted substances from a fluid, is applicable to reducing dissolved salt content in brackish and salt water. Cellulose acetate (CA) is a prime polymer used in RO desalination membranes and is the subject of this study (1).

Two factors are important in characterizing RO membranes.
1. The selectivity or ability to allow water to pass through while retarding other ion or molecule transport.
2. Flux or quantity of water transported through the membrane per area per time.

Reid and Breton (2) in the 1950's demonstrated that CA films had desirable separation qualities. However, the flux was not satisfactory. Subsequently, Loeb and Sourirajan (3), by varying the film casting technique, produced modified CA membranes. These membranes, cast from a two-phase swollen gel, developed a skin as they dried in air. Once the skin formed the solvent and swelling agent were leached from the gel leaving a porous substrate or an asymmetric membrane. With appropriate aqueous annealing (approximately 30 minutes at 85°C or 2 hrs. at 80°C), the dense skin provided good selectivity while the porous substrate provided structural strength without seriously hindering the flow. The properties of this type of a membrane are such that it can reject salt to the extent of $\sim$99%; and at a flux of $\sim$15 gal. flux/sq.ft./day at 1500 lb./in.^{2} (4). (Before flow occurs the pressure on the

307

feed water must exceed the osmotic pressure of the saline solution whether brackish or concentrated salt water.)

Besides the appealing simplicity of this desalination process, the energy requirements are low. Sharples (4) theorizes that the minimum energy to separate salt from sea water at room temperature is ~2.6kwh/100 U.S. gal. In RO many factors will increase this figure--pump efficiency, removal of concentration polarization (the increased osmotic pressure due to increased salt concentration immediately on the feed side of the membrane from which the product water is removed), and often the need to depressurize the liquid to atmospheric pressure by an inefficient process. Nevertheless, the energy requirement for RO will be less than half that for distillation of sea water. This advantage increases when the input is less concentrated. In addition, RO operates at ambient temperatures thus it avoids scaling and corrosion normally associated with saline solutions at elevated temperatures.

Two major features of RO membranes are not adequately clarified. First, the fundamental processes of water and ion transport through the membrane are not fully established. Secondly, the process known as compaction (the time rate of decrease in water flux through the membrane) which has been encountered in all types of polymer membranes is neither understood nor overcome.

Attempts to control compaction have resulted in the development of ultra-thin membranes by Lonsdale et al. (5). Ultra-thin membranes of dense CA correspond to the skin of asymmetric (Loeb) membranes. Using rigid, noncompressible porous substrates with ultra-thin membranes for selectivity also show compaction (6). Using high molecular weight CA or crosslinking to limit creep also proved unsuccessful. Compaction would appear to be a property of CA that requires a detailed investigation of the appropriate property/structure relationship.

The objectives of this investigation were:
1. To study in some detail the nature of internal relaxations in CA and in CA membranes, and to relate this information to the behavior of cellulose acetate RO membranes.
2. To investigate the effect of molecular weight, as well as of preparation conditions, and of subsequent annealing treatments on the physical behavior of CA film and membranes.
3. To investigate the presence of ordered regions in CA by use of x-ray diffraction and DSC and to correlate this with the results of the relaxation studies carried out by use of the TMA and with results of RO experiments.

4.   To investigate possible changes produced in the
structure and relaxation behavior of CA membranes
by continuous use as RO membranes and thus to improve our understanding of the compaction phenomenon.

EXPERIMENTAL

Various casting methods have been used to produce successful
cellulose acetate RO membranes.   Asymmetric membranes were considered representative of RO requirements for the present study.
In addition, dense films from acetone solutions were included in
this study to contrast with or eliminate the effects of the
porous substrate in asymmetric membranes.

Characterization of the Cellulose Acetate Materials Investigated

Several Eastman Kodak CA powders were utilized to produce
both dense and asymmetric membranes.   The main diacetate (partially substituted CTA) powders were the Eastman designation:
E398-3, E398-10, E394-45, and #394-60.   See Table 1.

A second group of films was obtained through the courtesy of
Dr. H. K. Lonsdale of Gulf General Atomic, Incorporated, San Diego,
California.   Tests were made on dense films of E398-3, E398-10,
and E394-45 that had been run for 7 days under normal operating
conditions and then removed from the RO test cell.   These were

TABLE 1

CHARACTERIZATION OF COMMERCIAL CELLULOSE ACETATE AND TRIACETATE IN THIS STUDY

Eastman Kodak desig- nation	Polymer form	Percent acetyl	Degree of substitu- tion ds a.	Commer- cial viscos- ity (seconds b.	Intrin- sic viscos- ity [η] c.	Degree of polymer- ization FP or P d.	Calculated molecular weight e.
398-3	Powder	39.8	2.438	3	1.01	131	35,100
398-10	Powder	39.8	2.438	10	1.28	166	44,800
394-45	Powder	39.4	2.425	45	1.63	212	57,300
394-60	Powder	39.4	2.425	60	1.69	220	59,400

a. Calculated from % acetyl content, where: ds = 164 x/100y-xz and 162 = formula
weight of one anhydrous glucosidic unit, x = % acetyl, y = formula weight of acetyl =
43, and z = net formula weight of acetyl-substituted hydrogen = 42.

b. The Eastman Kodak commercial viscosity designator is time in seconds for the ball-
drop test, ASTM, D 871-48, with a 20% acetone solution.

c. The intrinsic viscosity [η] is taken from a graph (23) which relates Eastman Kodak
commercial viscosity to intrinsic viscosity.

d· The degree of polymerization (DP or P) can be calculated from P = K[η] (23), where
for CA K = 130 if ds ~ 2.4 (23).

e. The molecular weights were calculated by DP x repeat unit weight = 270.

stored in water prior to mechanical testing. In addition, asymmetric membranes of proprietary CA material were tested in the compacted and noncompacted condition. Some of these were as-cast membranes, while others were water annealed. One group of samples was given a 70°C water anneal for 30 minutes and a second group was given an 85°C water anneal for the same length of time.

The membranes cast in our laboratory were prepared using various solvent systems. In the case of the dense acetone films, they were cast from a 25 wt.% acetone solution and are similar to the original Reid-Breton films. Other asymmetric films were cast by Loeb-Sourirajan procedures (8) using a water, acetone, and magnesium perchlorate formula, casting at −10°C, and quenching in ice water. The main asymmetric films studied were cast from a formamide-acetone solution being cast and quenched at room temperature. The formamide-acetone solutions tended to degrade. Hence, they were cast soon after preparation and the solutions were discarded after 24 hours.

Description of Film Casting Techniques

Films were cast by use of a 3-inch Multiple-clearance Film Applicator AG-3825N (Gardner Instruments, Inc.). Ordinary plate glass, from 6" x 10" to 12" x 24", was used for the substrate upon which the solutions were cast. Both dense and asymmetric films were cast on the glass, which was held at room temperature, or at −10°C for the perchlorate films. Distilled water or iced-distilled water was used for quenching and leaching when asymmetric films were made.

A constant temperature water bath was used for annealing films in water. A vacuum oven was used for annealing or drying a few specific films and powders. Wet films were stored in distilled water in covered beakers in the dark. Films were dried between a combination of bond paper and plate glass either in air or in a vacuum oven. The paper aided drying and the glass minimized deformations. The dried films were ambient stored in envelopes. Use of the term "dried" in conjunction with films refers to the ambient dry condition unless otherwise stated in this study. The films were of good quality with high optical clarity; few bubbles, if any, were present and the thickness, as measured by a micrometer, was uniform. The films were subsequently stored in distilled water or in flat envelopes.

It was felt that investigation of the modified films cast from the more swollen solutions would help to differentiate the mechanical properties of the RO films. Therefore, two types of asymmetric films were cast.

Asymmetric films formed when a dense surface layer developed due to evaporation and the solvent was leached from the remaining cast solution. This left the so-called porous substrate of rapidly coagulated polymer. Loeb-Sourirajan modified films with their high water flux were made from a solution of 22.2 wt.% CA (E398-3 was used in the original Loeb-Sourirajan films), 66.7 wt.% reagent grade acetone, 10.0 wt.% distilled water, and 1.1 wt.% Baker anhydrous magnesium perchlorate. Wet film thickness for all asymmetric membranes was standardized at the same 30 mils as for dense membranes. This gave films with the normal translucence common to acceptable asymmetric CA membranes. They were readily handled.

The Loeb solution was cast (in a freezer compartment) on plate glass which was at -10°C. After 30 seconds, the membrane was quenched in ice water for an hour. During the quenching, the film released from the plate glass. (The recommended 4-minute drying time gave a very opaque white film.) Their appearance indicated gross structural inhomogeneities. They had a fibrous texture but did not have visible bubbles.

Formamide-acetone asymmetric membranes are high flux membranes that can be cast and quenched at room temperature. The solution is 25 wt.% CA, 30 wt.% formamide, and 45 wt.% acetone. Films from various Eastman Kodak powders were cast and dried 30 seconds in air at room temperature. They were quenched in room temperature water for one hour. The appearance of these films was mildly translucent and very consistent. Few imperfections, if any, were evident.

Some of the membranes which were received from Gulf General Atomic, Inc., were reportedly prepared as dense membranes from 20 wt.% CA in acetone solutions on a moving substrate (9). They were dried to ambient conditions before being placed in the RO test cell.

Other films from Gulf General Atomic were asymmetric membranes. These were proprietary as to the CA component. The films were cast from a Loeb-Sourirajan formulation of 68 wt.% reagent grade acetone, 13.5% water, 1.5% magnesium perchlorate, and the rest CA. These were air-dried on the order of one minute before an ice-water quench.

The RO membranes from Gulf General Atomic were either as-cast films, films run in the test cell, and/or films that were heat-treated. The dense films, E398-3, E398-10, and E394-45, were received as-cast dried, and as-cast in the wet condition after being run in a test cell. Compaction, if present, resulted from pressurizing the membranes at 1500 psi with a 3.5% NaCl solution at 25°C for 7 days in the RO test cell. The asymmetric proprietary

membranes from Gulf General Atomic were each received wet in the
as-cast condition or with one of two heat treatments. The heat
treatments were 70°C or 85°C in water for 30 minutes. Each of
the three types of films had compacted counterparts.

Heat treatment of the dense and asymmetric films in this
laboratory consisted of vacuum oven annealing and annealing by
immersion in distilled water.

The vacuum oven annealing temperatures were selected on the
basis of differential scanning calorimetry. The DSC thermograms
revealed glass transition peaks between 185°C and 203°C for all
samples produced at our facilities. Consequently, all dried
samples were annealed between 203°C and 225°C for 2 hours.

The water-immersed annealing heat treatments were made at
70°C, 80°C, 90°C, and boiling for 2 hours. Corresponding to the
standard heat treatment, the 80°C treatment for 2 hours produced
the most noticeable change in x-ray diffractometer patterns. As
a consequence, the concluding studies of heat-treated films were
limited to samples which were heated at 80°C for 2 hours in water
for membranes cast in this laboratory.

Testing Procedures

X-ray Diffractometry. The use of wide-angle x-ray diffractometry with automated counter data presented in graphical form is
superior to a camera for parametric measurements when the diffraction lines are abnormally broad (10). For CA, this is certainly
the case. The diffractometer profiles of RO membranes are broad,
diffuse patterns (See Figure 3).

Wide-angle diffractometry appeared to offer the best resolution of changes in x-ray patterns due to thermal treatment or due
to compaction from being in an RO test cell.

The wide-angle x-ray diffractometer (11) used in this investigation is the North American-Philips System consisting of an
x-ray diffraction unit, Type 12045; a high-angle goniometer unit,
Type 42201; and an electronic circuit panel, Type 12048. The
precision of the angular measurements was $\sim 1/4$ of a degree.

Two types of scans were taken. First, single film thicknesses
were scanned in the ambient dry condition. Secondly, certain samples were observed as a sandwich of two films with distilled water
between. The films retained their wet condition during these runs.

Differential Scanning Calorimetry. The Model DSC-1B

Differential Scanning Calorimeter, manufactured by the Perkin-Elmer Corporation, measures the electrical energy required to maintain a zero temperature difference between the sample and a reference material (12). In this study, two references were used. Initially, the samples, both wet and dry, were run against an empty reference pan. Finally, wet samples were run against samples that were previously dried in the DSC at temperatures above 130°C. Both the sample and the reference were subject to a controlled temperature change.

The settings on the DSC can be varied as needed to give a good, discernible peak with a satisfactory base line and rate of temperature change. The sample size and moisture content affected these settings. The unit initially was calibrated with materials with known heating transitions. The resulting DSC thermograms describe a curve with peak areas directly proportional to the heat of the transformation (12). With a calibrated instrument, the DSC values were proportional to specimen heat capacity (13).

The samples of approximately 2 to 12 mg were sealed in weighed aluminum pans with covers which were closed. After a 5-minute wait for the test cell to be purged with a stream of purified and dried N_2, the test was started. The starting temperature depended on the speed and temperatures of interest, since one minute was required for the recorder to stabilize. The input heating rates studied were 10°C/min.

Thermal Mechanical Analyzer. Thermal mechanical analyzers measure both the expansion and softening of materials as a function of temperature. The materials studied were examined on the Perkin Elmer Corporation's Model TMS-1 TMA. Two modes of operation were available. First, the expansion mode measured dimensional changes. The TMA can indicate heat distortion temperature, first- and second-order transitions, creep, specific volume changes, and coefficient of linear thermal expansion as well as glass transitions over a broad temperature range. The TMA can give coefficients of expansion under various loads (14).

The main use of the TMA in this study was for viscoelastic analysis. Two load levels were used. The so-called penetration probe was used with a 4-gram load and the expansion or flat probe was used with a 200-gram load (Figure 1). The 200-gram load eliminated the effects of the film warping. Penetration is the most sensitive mode of operation; it is best for measurements of softening or melting. Expansion gives the most conclusive data for glass transition.

The samples were run at 10°C/min. heating rates with a dry helium purging gas flow. They were tested in DSC sample pans in

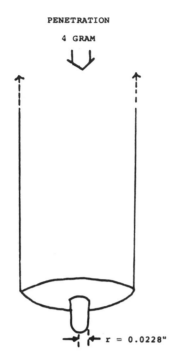
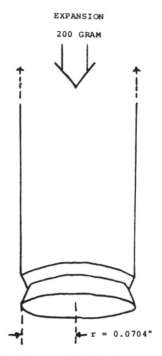

PENETRATION   EXPANSION

4 GRAM   200 GRAM

r = 0.0228"   r = 0.0704"

Radius = 0.0228 in.   Radius = 0.0704 in.
Area = 0.001632 in.^{2} Area = 0.01557 in.^{2}
Load = 4 gms. = 0.0088 lbs.   Load = 200 gms. = 0.440 lbs.
Pressure = L/A = 5.4 psi   Pressure = L/A = 28.3 psi

Figure 1.   Probe and loading diagram.

the ambient dry condition, swabbed dry, and with the pan filled
with water. The samples were punched from the films to fit the
aluminum sample pans.

Empty aluminum sample pans demonstrated no significant change
in thickness over the near 0°C to 280°C range studied. The quartz
probes were cleaned between each run with acetone and glacial
acetic acid.

RESULTS AND DISCUSSION

CA may not form true crystalline regions in the definitions
in crystallography. CA usually denotes a partially substituted
triester (e.g., where all of the hydroxyl groups which may be
substituted are not replaced by acetyl groups). Therefore chemical periodicity cannot be anticipated for randomly substituted CA.

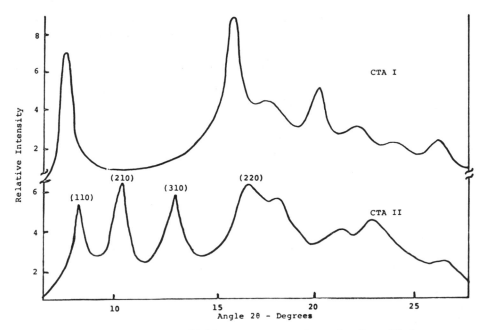

Figure 2a. X-ray diffractograms of typical cellulose
triacetate I and II (crystalline).

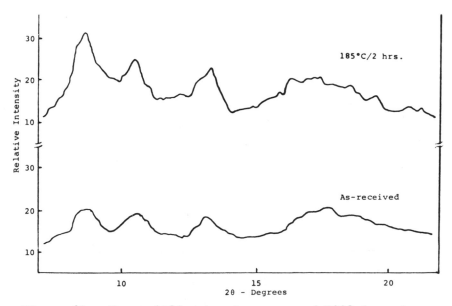

Figure 2b. X-ray diffractometer scans of E398-3 powder
in as-received and annealed conditions.

However, from an engineering viewpoint, if the mechanical properties of a polymer can be influenced by ordered aggregates which
act as physical tie-points, the material may be considered "quasi-
crystalline" or ordered. In this investigation the term "crystalline", when applied to CA, will define regions that produce endothermic peaks as determined by Differential Scanning Calorimetry
(DSC). The DSC endothermic peak is proportional to the heat of
fusion of the material. In this case, the randomly substituted
side groups may be considered as local defects in an otherwise
crystalline array of the basic glucosidic repeat units in the CA
backbone.

Wide-angle x-ray diffraction profiles agree since only films
cast from CTA powder could be made to produce a crystalline type
pattern reported by Watanabe, Takai, and Hayashi (15). See
Figure 2a. However, in one CA material, the low molecular weight
E398-3 powder did show the development of some crystalline regions
when annealed in air. See Figure 2b.

Commercial Eastman Kodak powders E398-3, E398-10, and
E394-60 were studied by x-ray diffractometry plus differential
scanning calorimetry to observe changes in as-received and annealed conditions. X-ray detectable order occurred in E398-3
above 185°C while minor changes in order occurred in water above
70°C. In diffraction profiles from films cast of either CA or
CTA, the two diffuse reflections common to amorphous polymers were
present. See Figure 3. The amorphous pattern altered showing
improved order. The change in CA diffraction patterns did not
indicate the degree of order present in patterns of annealed CTA.

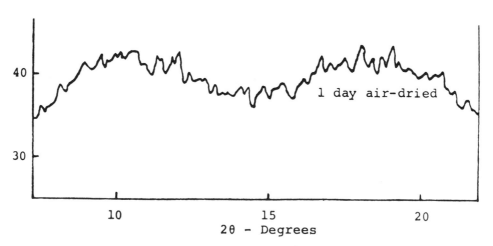

Figure 3.   X-ray diffractometer scan of E394-60
            acetone cast membranes.

Starting with wet films of CA and making sequential diffraction patterns with drying indicated structural changes. In the case of as-cast acetone cast E398-3 films, drying showed that the lower 2θ peak, in its amorphous type profile, increased in size and appeared to move to a lower 2θ angle. However, for similar E394-60 films, drying showed the higher 2θ peak increases. The water in RO cellulose acetate membranes was structurally integral and its movement in the film could change molecular arrangements.

Further structural details of CA were obtained with DSC. Crystalline type melting endotherms were indicated for certain CA materials, even though crystallographic, three-dimensional periodicity was impossible in randomly substituted CA. Of the as-received Eastman Kodak CA powders, only E398-3 showed significant crystalline order in sample weights of 2 mg. Quenching E398-3 from above the melting temperature of 230°C nearly eliminated the melting endotherm, but annealing near or above the primary glass transition (190°C) caused recovery of the melting endotherm.

For dense CA membranes cast from acetone solutions, only E398-3 films displayed significant crystallinity in 2 mg sample. See Figure 4. However, 10-15 mg samples of E398-10, E394-45, and E394-60 showed a melting endotherm at around 240°C. The presence of more crystallinity in the lower molecular weight E398-3 could be expected. The higher melting temperature normally considered to represent large crystallite units was unexpected in the higher molecular weight material.

Asymmetric membranes from formamide-acetone solutions showed melting endotherms at 230°C for 2 mg samples of E398-3 and E398-10. See Figure 5. Therefore, the E398-10 membranes from the formamide-acetone solution appeared to develop more crystallinity. In the case of the E398-10, the melting temperature was also in the 230°C range.

In DSC profiles of wet versus dry membranes, two transitions occurred below the T_m and T_g. The one designated as T_1 at 50°C was an exotherm representative of ordering. The other was the T_2 which was a shift in the base line and occurred at 120°C to 135°C. This was preceded by an endotherm between 70°C and 103°C, which represented water leaving the membranes.

Dense CA membranes from acetone solutions were investigated with a Differential Scanning Calorimeter (DSC), and Thermal Mechanical Analyzer (TMA). Observed alterations in diffraction patterns of dense acetone films indicated changes due to molecular weight and specific thermal treatments.

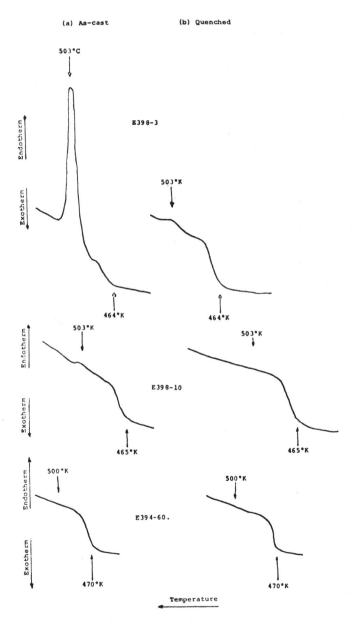

Figure 4. DSC thermograms for (acetone) cast films.

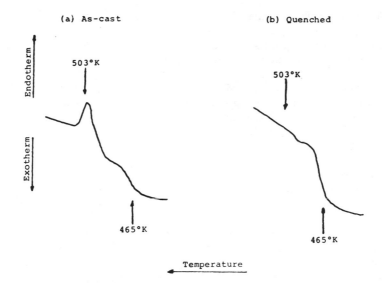

Figure 5a. DSC thermograms for (acetone/formamide) cast film
of E398-10.

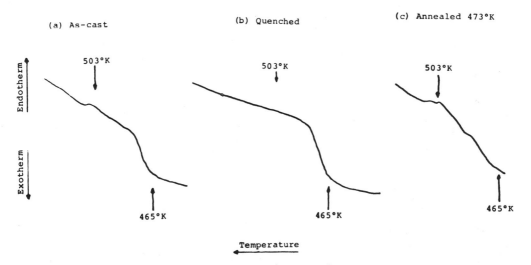

Figure 5b. DSC thermograms for (acetone) cast film of E398-10.

TMA results for dense CA membranes exhibit three transitions between room temperature and the final melting (softening) transition (T_m). See Figures 6 and 7 and Table 2. The T_1 transition is fairly broad and starts near room temperature. It is relatively insensitive to water content and appears as a DSC endotherm (normally associated with ordering). The T_2 transition near 120°C has a relatively narrow range and requires the presence of water. The T_2 temperature is lowered by annealing in water but thickness change (TMA) is increased. Annealing appears to alter the distribution of water (swelling). The primary glass transition (T_g) starts near 190°C. For acetone cast E398-3 films, the T_g cannot readily be separated from the T_m because low molecular weight CA has a reduced melting transition. In E398-10, E394-45, and E394-60 membranes, T_m is around 240°C.

The TMA data is presented as line drawings of tangents of the actual curves to readily show the transitions. The transition temperatures in the tables are the midpoints of regions where significant differences in thickness are occurring. Table 3 is a compilation of transitions reported in the literature of CA and CTA.

"Loeb" type or asymmetric membranes cast from acetone-formamide were studied with the TMA. Four transitions, including the T_m found in dense membranes, were present in asymmetric membranes, generally occurring at the same temperatures as in dense membranes and are similarly affected by annealing in water. See Figure 8 and Table 4. The most significant difference occurs in the increased thickness change associated with T_2, which is directly related to water leaving the membrane. The monitoring thermo-couple indicates evaporative cooling during T_2. The large reduction in thickness associated with T_2 must include collapse of the porous substrate.

Gulf General Atomic Laboratories supplied dense membranes, pseudo-compacted for one week in a reverse osmosis test cell, and proprietary "Loeb" type membranes in as-cast, heat-treated, compacted as-cast, and compacted heat-treated conditions. See Tables 5 and 6.

TMA shows four transitions present in all films except dense E398-3 films where the T_g and T_m overlap. See Figure 9 and Table 7. The asymmetric membranes show that a 70°C/30 min. heat treatment in water reduces nominal, as-cast thickness while 85°C/30 min. produces more swollen or structurally stronger membranes. The T_2 thickness changes in noncompacted heat-treated membranes are extensive. See Figure 10. Compacted heat-treated asymmetric membrane thickness is significantly reduced. See Figure 11. T_2 thickness change is greatly reduced. The porous substrate severely collapses due to compaction. The noncompacted and

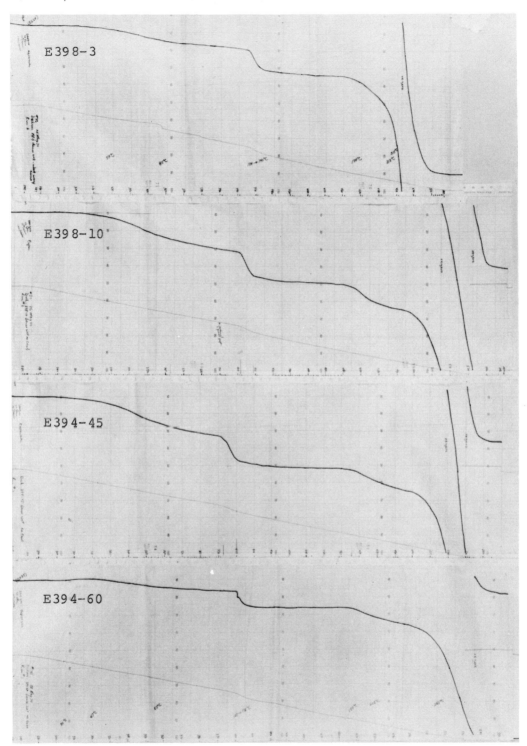

Figure 6.   Actual TMA scans.

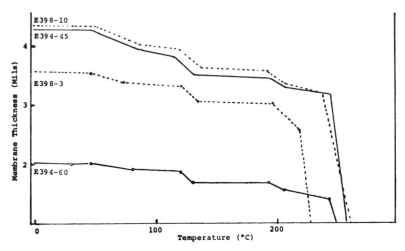

Figure 7.   Dense, wet E398-3, E398-10, E394-45, and E394-60 films
(10°C/min. heating rate).

TABLE 2

DENSE, WET MEMBRANES--FLAT PROBE--200-GRAM LOAD

Material	Film condition	Transition temperatures (°C)			
		T_1	T_2	T_g	T_m
E398-3 25 wt.%	As cast	61	128	196+	219+
E398-10 25 wt.%	As cast	69	128	19⁰	237+
E394-45 25 wt.%	As cast	65	123	200	244+
E394-60 25 wt.%	As cast	63	125	200	243+
E398-3 20 wt.%	Run in RO test cell	51	125	208+	222+
E398-10 20 wt.%	Run in RO test cell	56	127	199+	244+
E394-45 20 wt.%	Run in RO test cell	49	119	196+	240+
E398-3 25 wt.%	80°C/2 hrs. water	45+	120	194	218+
E398-10 25 wt.%	80°C/2 hrs. water	56+	120	200	235+

TABLE 3

SOME REPORTED TRANSITION TEMPERATURES IN CELLULOSE ACETATE

Material	Method	Transition temperatures (°C)				Reference
		T_1	T_2	T_g	T_m	
Cellulose triacetate	DSC	47	?	?	285	(16)
Cellulose triacetate	Specific volume	40	120	155	280-300	(17)
Cellulose triacetate	DTA and TBA	None	None	190	290-309	(18)
Cellulose triacetate	Specific volume	30	105			(19)
Cellulose triacetate	Specific volume	46	112	157		(20)
Cellulose triacetate	Specific volume	30	105			(21)
Cellulose triacetate	Dynamic mechanical	--	105	175		(17)
Cellulose triacetate	Dynamic mechanical	60	--	180		(21)
Cellulose acetate (2.5)	DSC	55	?	?	230	(16)
Cellulose acetate (2.2)	Specific volume	55	115		None	(17)
Cellulose acetate (2.5)	Specific volume	60	120			(19)
Cellulose acetate	Specific volume			68.6		(22)
Cellulose acetate (2.5)	Specific volume	48	118			(23)
Cellulose acetate (2.5)	Ionic conductivity	40-60	80-100			(23)
Cellulose acetate (2.2)	Dynamic mechanical	--	130	195		(17)
Cellulose acetate	Dynamic mechanical		120			(21)

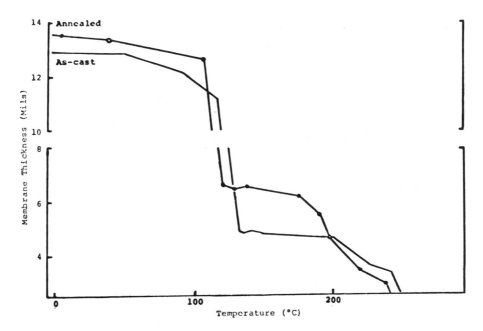

Figure 8. Asymmetric-formamide-acetone, wet E398-10 membranes
(as-cast vs. 80°C/2 hrs. water annealed)(10°C/min. heating rate).

TABLE 4

ASYMMETRIC E398-10 MEMBRANES

Film condition		Test condition		Transition temperatures (°C)			
Solvent	Heat treatment	Moisture	Probe	T_1	T_2	T_g	T_m
Formamide		Wet	Heavy	72	126	214	244+
Formamide	80°C/2 hrs. water	Wet	Heavy	40+	115	196	240+

TABLE 5

COMPACTION DATA FOR DENSE FILMS

Elapsed time (hours)	Membrane	Flow (10^{-3} cc/min.)	Rejection (percent)
83	398-3, 35 μ	2.29	98.7
127	398-3, 35 μ	2.32	99.5
154	398-3, 35 μ	2.98	99.6
83	398-3, 35 μ	1.89	86.5
127	398-3, 35 μ	1.98	94.4
154	398-3, 35 μ	No data	No data
83	398-10, 50 μ	2.04	87.9
127	398-10, 50 μ	2.15	97.3
154	398-10, 50 μ	2.67	98.9
83	398-10, 50 μ	0.65 (?)	97.1
127	398-10, 50 μ	2.57	98.0
154	398-10, 50 μ	2.56	98.6
103	394-45, 40 μ	2.68	97.8
127	394-45, 40 μ	2.70	99.0
154	394-45, 40 μ	3.47	99.0

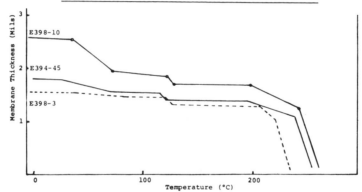

Figure 9. Gulf dense, wet, test cell run E398-3, E398-10, and E394-45 membranes (10°C/min. heating rate).

TABLE 6

COMPACTION DATA FOR "LOEB" TYPE FILMS

Elapsed time (hours)	Annealing temperature (°C)	Flow (cc/min.)	Rejection (percent)
72	25	3.2	22
95	25	3.4	22
136	25	3.1	25
166	25	3.0	28
72	25	0.78 (?)	28
95	25	0.74 (?)	29
136	25	0.58 (?)	35
166	25	0.35 (?)	42
72	70	1.2	88
95	70	1.3	88
136	70	1.2	87.3
166	70	1.0	88

TABLE 6 (cont'd)

COMPACTION DATA FOR "LOEB" TYPE FILMS

Elapsed time (hours)	Annealing temperature (°C)	Flow (cc/min.)	Rejection (percent)
72	70	1.2	86
95	70	1.3	86
136	70	1.2	87
166	70	1.0	88
72	85	0.68	97.1
95	85	0.70	97.1
136	85	0.65	97.1
166	85	0.56	97.1
72	85	0.72	97.1
95	85	0.75	97.1
136	85	0.68	97.2
166	85	0.60	97.2

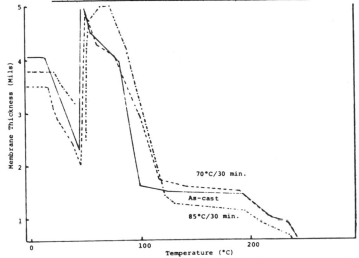

Figure 10. Gulf asymmetric, wet membranes (as-cast vs. 70°C/min. and 85°C/30 min. water annealed)(light probe--10°C/min. heating rate).

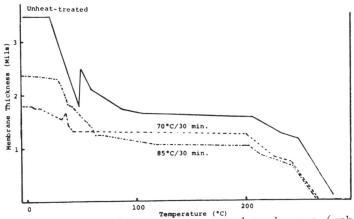

Figure 11. Gulf asymmetric, wet, compacted membranes (unheat-treated vs. water annealed)(light probe--10°C/min. heating rate).

compacted as-cast Gulf membranes behave similarly. Here T_2
thickness changes are less than non-compacted heat-treated films,
while T_1 thickness changes are significantly greater.

Use of the TMA where testing could be conducted in water
revealed initial thickness differences in dense type films. With
this technique, the wet films of E398-10 and E394-45 were thickest.
In E398-10 and E394-45 membranes, the change in thickness of the
T_1 was also greater. Similarly cast films from Gulf General
Atomic Inc., which were subjected to 7 days of 1500 psi operating
pressure in RO test cells also showed the same rank in original
thickness. See Figure 9. Because of the low water flux through
these dense type membranes, they were considered to be in a pseudo-
compacted state (i.e., an initial stage of compaction when flux
is actually increasing). Utilizing these Gulf films for effects
of compaction was very limited. Annealing E398-3 and E398-10
dense films in 80°C water for 2 hours resulted in films of similar
thickness to each other. The E398-3 became more swollen while the
E398-10 became less swollen. Annealing E398-3 films may have
affected the crystallite tie-points by swelling them. In the case
of E398-10, annealing which caused x-ray-detectable order reduced
the thickness by ordering some of the amorphous regions.

The TMA data from proprietary asymmetric membranes supplied
by Gulf General Atomic Inc. are presented in Figures 10, 11, 12
and 13 and Table 7.

Figure 10, for the lightly loaded non-compacted samples, indicates that heat treatment affects their overall thickness. The
70°C/30 min. in water annealing causes twice the reduction from
the original thickness of the untreated films, as does the 85°C/
30 min. in water annealing. Beyond this, the T_1 transition in
these films begins below room temperature except for those
annealed at 85°C/30 min. in water. The T_1 transition in all of
these films ends abruptly with a large apparent expansion and a
marked warping of the film. Once the film is buckled, it relaxes showing a T_2 type transition, but at temperatures lower than
the T_2 transitions common to those resolved with the flat probe
under the 200-gram load. The T_g and T_m transitions are similar
for all heat-treated conditions.

When the same films are compacted (e.g., run in an RO test
cell), the same testing conditions produce the thermograms in
Figure 11. Comparing Figures 10 and 11 shows that the probe is
settling on thinner membranes once they have been placed in the
test cell. The as-cast films are either less affected by compaction, the annealed films cannot recover as well after being
removed from the test cell, or residual stresses are partly removed by annealing and partly by time in use.

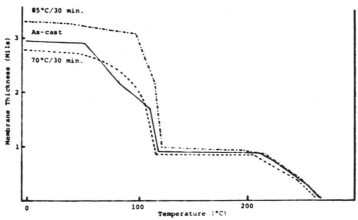

Figure 12. Gulf asymmetric, wet membranes (as-cast vs. water annealed) (heavy probe--10°C/min. heating rate).

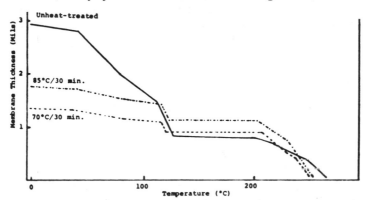

Figure 13. Gulf asymmetric, wet, compacted membranes (unheat-treated vs. water annealed)(heavy probe--10°C/min. heating rate).

TABLE 7

WET, ASYMMETRIC GULF GENERAL ATOMIC PROPRIETARY MEMBRANES

Film characterization		Test condition	Transition temperatures (°C)				
Heat treatment	Condition	Probe	T_1	E	T_2	T_g	T_m
Unheat treated	Not compacted	Light	30	48	92	207	235+
70°C/30 min.	Not compacted	Light	32	49	97	203	236+
85°C/30 min.	Not compacted	Light	38	54	99	204	238+
Unheat treated	Compacted	Light	37	51	69	220	249+
70°C/30 min.	Compacted	Light	21	33	41	214	243+
85°C/30 min.	Compacted	Light	35	40	53	208	244+
Unheat treated	Not compacted	Heavy	69		115	217	252+
70°C/30 min.	Not compacted	Heavy	*		112	205+	245+
85°C/30 min.	Not compacted	Heavy	42+		109	226	247+
Unheat treated	Compacted	Heavy	61		120	211	249+
70°C/30 min.	Compacted	Heavy	59		118	219	239+
85°C/30 min.	Compacted	Heavy	61		121	205+	232+

*These membranes drift into T_2. No definite T_1 is present. E is the indicated expansion due to buckling of membranes.

There are four transitions in all of these films, plus a
"buckling" phenomenon. The buckling occurs approximately at the
same temperature range as in the non-compacted films. However,
the extent of warping is considerably less and, in the case of
the specimen annealed at 85°C/30 min. in water, no net expansion
is observed. There is only a break in the slope between the T_1
and T_2 transitions but no net rise in thickness. In fact, the
two transitions (T_1 and T_2) reported for these compacted films
may be the previously reported T_1 transition with the warping
phenomenon superimposed. At high temperatures T_g and T_m are
present. One fact to be noted is that at these temperatures
glass transition takes place over a small range of temperature
for the 85°C/30 min. films. The final softening or "melting"
temperatures are higher for the compacted films.

The entire series of heat-treated, compacted, and non-
compacted films was tested using the flat probe with the 200-gram
load. The thermograms are presented in Figures 12 and 13. With
the greater loading, the probe initially settles at lower thickness, indicating some collapse of the films. The as-cast, unheat-
treated film is most prone to this phenomenon. The annealed films
show the better structural integrity. As occurred in other annealed films, the T_1 transition is affected, resulting in no
definite temperature at which marked changes occur for the 70°C/
30 min. films. In fact, the probe only indicates a general drift
until the T_2 transition occurs. In the 85°C/30 min. films, the
start of T_1 is evident but does not end before T_2 occurs. The
changes in thickness of these films are slightly higher than for
the round probe-investigated films, and the 70°C/30 min. films
have T_m start before T_g is complete. The softening or "melting"
point is also higher for these films. The values obtained appear
to be closer to those for the compacted films.

When these films are compacted and scanned with the flat
probe and the 200-gram load, the annealed films show a further
reduction in thickness (see Figure 12). Compared to the non-
compacted state, those films tested with the round probe and light
load are about one-half their original thickness. Only the unheat-
treated films appear to be capable of resisting this effect.
After compaction, the four transitions reported for other CA membranes are also present in these three types of films. These
transitions are listed in Table 7. Other than the changes in
thickness associated with each transition, they are similar in
temperature position. One exception may be the depressing in the
final softening or melting point of the compacted 85°C/30 min.
films. At temperatures above about 120°C, the annealed films
seem to resist the probe's penetration better than the unheat-
treated films.

CONCLUSIONS

In Gulf General Atomic proprietary asymmetric membranes, the temperature of the annealing water affected the original thickness as determined by the point where the probe initially came to rest. The 85°C water annealing produced thicker membranes than those annealed in 70°C water which were thinner than the as-cast membranes. With the heavier flat probe, the 85°C water annealed films were thicker than either the as-cast or 70°C water annealed films, which might indicate that they were structurally stronger.

All wet membranes that were tested demonstrated four TMA transitions--T_1, T_2, T_g and T_m--except dense E398-10 films cast from acetic acid where T_g and T_m were superimposed. With a midpoint temperature between 50°C and 70°C, the T_1 transition was present in wet and dry membranes. In dry films, the thickness change associated with T_1 was reduced. Annealing the wet films reduced the thickness change and moved the disperse T_1 transition toward and into the lowered T_2 transition. Solvent and molecular weight did not affect the T_1 temperature. Light loading in asymmetric membranes showed that this transition starts near room temperature.

This transition is speculated to be associated with segmental motion of the main chain, which may be accompanied by acetyl motion in amorphous regions, and the movement of water. Bond rearrangement would occur, altering the order of the polymer, and when water is present it alters the number of water molecules bridging between hydrogen bonding sites in the amorphous regions.

The stability of bonding arrangement in the amorphous region, as indicated by changes in the T_1 transition based upon TMA results, is affected by loading, quenching temperature, annealing, and compaction. The heavily loaded flat probe raises the start of T_1, possibly because the water cannot leave the membrane readily. Quenching in ice water may lower the T_1 temperature range by locking the main chain segments into less stable positions, while annealing above the T_1 in water allows ordering (DSC endotherm) of these segments. Annealing also raises the T_1 temperature range, indicating that the segments are less free to move perhaps due to new ordered tie-points. On the other hand, compaction, which reduces membrane water content, also lowers the T_1 temperature in as-cast and annealed films. Thus the stability of the segments is also affected by water in the membranes.

The effect of water was more noticeable in the T_2 transition which occurred between 110°C and 135°C for wet samples tested with the heavily loaded flat TMA probe. No transition occurred in ambient-dry films. In wet films during the transition the exit of water was indicated by evaporative cooling of a monitoring thermo-couple. The transition occurred over a narrow temperature range and the thickness change was significant, especially in asymmetric membranes where obvious participation of the porous substrate was involved.

The T_2 transition is believed to be the result of a change in ordered polymer regions involving the number of water molecules in bridging between hydrogen bonding sites (i.e., the degree of swelling in ordered regions). The narrow temperature range of the T_2 transition indicated a limited range of relaxation times or activation energies. The temperature of the transition was influenced by molecular weight, drying, and resoaking pseudo versus real compaction and annealing.

In E394-45 and E394-60 films the T_2 transition is lower than in the lower molecular weight films. Apparently, these originally less ordered films can swell and allow more water molecules per bridge. Drying to ambient condition followed by resoaking for one hour raises the T_2 temperature. Presumably, in the one-hour soak fewer water molecules per bridge are involved. In addition, the T_2 thickness change for resoaked films is less. In pseudo-compacted dense membranes, the T_2 transition temperature is lowered. Where real compaction has occurred in asymmetric films, the T_2 temperature is raised. Annealing reduces the temperature of T_2 and increases the thickness change. Annealing appears to cause an interaction between the material involved in the T_1 and T_2 transition. The thickness change of the T_1 transition is reduced while the T_2 transition is accompanied by a greater thickness change. This may result from the segmental motion which would occur during annealing above T_1. The motion would allow amorphous regions to order and the water to form bridges in the new and previously ordered regions. The difference in swelling of asymmetric films annealed at 70°C versus 85°C indicates that the degree of swelling is temperature dependent. The extent of the T_2 thickness change is sufficient to require the involvement of the porous layer. Consequently, the water in this region must hypothetically be in a bridging or bound configuration.

The primary glass transition, T_g, was present and occurred at slightly higher temperatures than in the DSC. It was not particularly molecular weight sensitive, nor was it affected by annealing. The transition occurred at slightly higher temperatures in asymmetric membranes.

The T_g is associated with gross molecular slipping of polymers in the amorphous regions.

The final TMA transition was the melting or final softening, which was molecular weight sensitive. The E398-3 membranes melted before the higher molecular weight materials. Membranes cast from acetic acid had particularly low melting temperatures. Hypothetically, these low melting temperatures may arise from small folded-chain crystallites in low molecular weight material or slow setting films cast from solutions with low vapor pressures. In the other films, larger micellar type crystallites may occur.

Annealing in water did not affect the T_m, except in Gulf asymmetric membranes where increased swelling appeared to decrease the order. However, asymmetric films which severely warped (buckled) also had their T_m's reduced in temperature. The warping appeared to cause a reduction in order in the membranes.

"Buckling" is a phenomenon associated with asymmetric membranes. It is presumed to result from releasing internal stresses introduced by quenching and leaching of solvent from the porous layer. If the sample is lightly loaded, the buckling can be severe and occurs during the latter part of the T_1 transition. In some heavily loaded samples it shows a minor change after the T_2 transition. Both annealing and compaction appear to reduce the phenomenon; therefore, it appears to involve bond rearrangement.

Compaction, another phenomenon in RO membranes, appears to have two stages. Initially, as water passes through the membrane, flux increases. This appears to occur when bonds are being redistributed due to segmental motion at room temperature. The second stage occurs when these segments become ordered. The more regular polymer network developed allows bound or more ordered water to occur between hydrogen bonding sites. As in the concept of Testa and Bruins (24), selectivity would increase. The more stable hydrogen bonded water would likely diffuse more slowly, resulting in the lower flux of the second stage of compaction. The idea that the initial stage of compaction affects molecular arrangement is demonstrated in the T_g of the pseudo-compacted dense Gulf membranes. The T_g is not complete before the T_m occurs, while in the as-cast films this only occurs in E398-3 membranes.

The presence of T_1 and T_2 seems integrally connected with the effects of annealing, warping, and compaction in RO membranes. Since these transitions have been associated hypothetically with segmental motion resulting in ordered regions containing bridging water molecules, x-ray diffraction patterns of wet Gulf asymmetric membranes were made.

The wide-angle profiles indicated that annealing enhanced the amorphous type pattern near the (010) interplanar spacing. Compaction greatly reinforced this peak. Thus ordering appears to occur.

As a speculation, the ordering is, at least, some form of planar array of molecules in the plane of the film. If the glucosidic rings lying in the plane of the film form these arrays, then a system similar to graphitic oxide membranes would exist in the CA membranes. The orientation of the glucosidic rings, which would produce a more tortuous path, and the reduction in water molecules bridging between rings would lower flux and increase selectivity.

REFERENCES

1. Dickson, J., "Thermal Mechanical Properties of Cellulose Acetate Films and Asymmetric Membranes," Partial fulfillment of Ph.D. requirements, Rutgers – The State University of New Jersey, January, 1971.
2. Reid, C. E. and E. J. Breton, "Water and Ion Flow Across Cellulosic Membranes," Journal of Applied Polymer Science, I, 133-143 (1959).
3. Loeb, S. and S. Sourirajan, "Sea Water Demineralization by Means of a Semipermeable Membrane," UCLA Report, 60-60 (July, 1960).
4. Sharples, A., "An Introduction to Reverse Osmosis," Chemistry and Industry, 322 (March 7, 1970).
5. Lonsdale, H. K., U. Merton, R. L. Riley, and K. D. Vos, "Reverse Osmosis for Water Desalination," Research and Development Progress Report No. 208, to OSW, General Dynamics, General Atomic Division (Sept., 1966).
6. Lonsdale, H. K., R. L. Riley, C. R. Lyons, and U. Merten, "Preparation of Ultrathin Reverse Osmosis Membranes and the Attainment of Theoretical Salt Rejection," Journal of Applied Polymer Science, II, 2143-2158 (1967).
7. Ott, E., H. M. Spurlin, and M. W. Grafflin, Cellulose and Cellulose Derivatives, Parts I and II, Interscience Publishers, Inc., New York (1954).
8. Loeb, S. and S. Sourirajan, "Sea Water Demineralization by Means of an Osmotic Membrane," Advan. Chem. Ser., 38, 117 (1963).
9. Lonsdale, H. K., Personal communication (1970).
10. Cullity, B. D., Elements of X-ray Diffraction, Addison-Wesley Publishing Co., Inc., Reading, Mass. (1959).
11. "X-ray Diffraction Instruction and Operating Manual," North American Phillips Co., Inc., Mt. Vernon, N.Y.

12. "Operating Instructions--Differential Scanning Calorimeter," Perkin-Elmer Corp., Norwalk, Conn. (1968).
13. Foltz, C. R. and P. V. McKinney, "Quantitative Study of the Annealing of Poly(Vinyl Chloride) Near the Glass Transition," Journal of Applied Polymer Science, 13, 2235 (1969).
14. "Operating Instructions for TMS-1 Thermal Mechanical Analyzer," Perkin-Elmer Corp., Norwalk, Conn.
15. Watanabe, S., M. Takai, and J. Hayashi, "An X-ray Study of Cellulose Triacetate," Journal of Polymer Science: Part C, No. 23, 825-835 (1968).
16. Kokta, B., P. Luner, and R. Suen, Quarterly Report to OSW on Contract 14-01-0001-1263 (Aug., 1968).
17. Russell, J. and R. C. Van Kerpel, "Transitions in Plasticized and Unplasticized Cellulose Acetates," Journal of Polymer Science, 25, 77-96 (1957).
18. Gilham, J. K. and R. F. Schwenker, Jr., "Thermomechanical and Thermal Analysis of Fiberforming Polymers," Appl. Polym. Sym. #2, 59-75 (1966).
19. Mandelkern, L. and P. J. Flory, "Melting and Glossy State Transitions in Cellulose Esters and Their Mixtures with Diluents," J. Am. Chem. Soc., 73, 3206 (1951).
20. Sharples, A. and F. L. Swinton, "Second-Order Transitions in Solutions of Cellulose Triacetate," Journal of Polymer Science, 50, 53-64 (1961).
21. Nakamura, K., "Studies on Viscoelasticity of Cellulose Derivative Films, Part I," Chemistry of High Polymers (Japan), 13 (130), 47 (1956).
22. Clash, R. F., Jr. and L. M. Rynkiewicz, "Thermal Expansion Properties of Plastic Materials," Ind. Eng. Chem., 36, 279 (1944).
23. Barker, R. E., Jr. and C. R. Thomas, "Glass Transition and Ionic Conductivity in Cellulose Acetate," Journal of Applied Physics, 35(1), 87-94 (Jan., 1964).
24. Testa, L. A. and P. F. Bruins, "Cellulose Acetate Membranes for Waste Water Purification," Modern Plastics, 45(9), 141 (May, 1968).

MECHANICAL PROPERTIES OF POLYMERS AT CRYOGENIC TEMPERATURES

A. Hiltner, J. R. Kastelic and E. Baer

Division of Macromolecular Science, Case Western

Reserve University, Cleveland ,Ohio 44106

SUMMARY
The relaxation behavior of linear polymers below 80°K is
reviewed. Tensile behavior over the temperature range 4.2-300°K
is reported for one linear polymer, polyethylene terephthalate.
Three regimes in tensile behavior are distinguished, and
transitions from one regime to the next are correlated with the
onset of molecular relaxation processes. The δ_c process at about
50°K is associated with departures from linearity in the stress
strain curves. The transition from a temperature dependent
fracture stress to a temperature independent fracture stress also
occurs at the δ_c temperature. The transition from brittle to
ductile behavior is observed in the temperature region of the γ
process, 150-200°K. A fourth behavior regime (45-140°K) can be
observed when specimens are tested in the presence of nitrogen.
Crazing and apparent yielding characterize the behavior within this
region.

BACKGROUND
Most polymers exhibit secondary relaxation processes below the
glass transition or melting point. These are usually associated
with the loosening of molecular constraints and the onset of
restricted motion of either side groups or short backbone segments.
Processes operative at the lowest temperatures are generally the
most primitive. It is reasonable to assume that processes which
become active at higher temperatures do not occur independently,
but are built up on the lower temperature modes by involvement of
larger units of the molecule in more complex motion. An understanding of the lowest temperature relaxation processes is thus of
fundamental importance to the interpretation of higher temperature
relaxations.

Dynamic mechanical relaxations in polymers below 80°K have been reported by various authors. The most extensively studied cryogenic relaxation is the δ peak in polystyrene (PS) at about 40°K (1cps). Data obtained by both dielectric and mechanical techniques for a wide variety of substituted polystyrenes can best be interpreted by considering possible motions of a phenyl group located within a cavity. The broad spectrum of observed relaxation times implies that the conformation of neighboring chains in the glass varies considerably to produce cavities of many different sizes and shapes. Cryogenic peaks observed in other polymers with side groups have been attributed to motions of pendant methyl or ethyl groups. These cryogenic side group processes have been the subject of several recent review articles[1,2].

The possibility that linear, crystalline polymers might show relaxation processes below 80°K has received less attention. Early investigators reported that polyethylene, polytetrafluoroethylene, and nylon 66 are viscoelastically inactive below 80°K[3,4]. However with the use of a very sensitive torsional pendulum, very low temperature relaxations have been observed in this laboratory in a number of linear polymers. Part of this paper will be concerned with the experimental features of cryogenic relaxations in linear polymers, and the extent to which these relaxations can be identified with motion of specific molecular entities.

The relation between relaxations and fracture is not well understood, and the occurrence of rate-dependent micromechanical processes at cryogenic temperatures raises the question of how the relaxations are revealed in macromechanical phenomena such as fracture. Results of Roe and Baer[5] demonstrated a correlation between the temperature of the γ relaxation and the transition from brittle to ductile tensile behavior for a number of linear polymers. This brittle-ductile transition usually occurs in the temperature range of 100-200°K. Some recent work will be presented here which indicates that certain features of the tensile behavior at temperatures below 80°K might also be associated with cryogenic relaxation processes.

EXPERIMENTAL

Dynamic mechanical measurements were made with an inverted torsional pendulum (1cps) which has been described previously[6]. The instrument was specifically designed for detection of low loss levels ($\Delta \times 10^{-4}$) at cryogenic temperatures using thin film specimens. The apparatus was evacuated for 24 hours prior to cooling, and measurements were made with the specimen under a helium atmosphere.

For the tensile tests, specimens of modified ASTM D412-68 Die C geometry were cut from 5 mil sheets of amorphous unoriented polyethylene terephthalate (PET) and polycarbonate (PC). The PET films were the same as used for the relaxation measurements. The method of cutting, utilizing a double template cutting guide and razor knife, yielded specimens free of nicks and defects suitable for low temperature tensile studies. These were tested in tension at one percent per minute strain rate throughout the temperature range 4.2 to 300°K using Instron mounted equipment similar to that described previously[5]. Appropriate seals and vacuum fittings enabled tests to be carried out in a mechanical vacuum or a controlled gasious atmosphere. Those specimens tested in vacuum were degassed in the apparatus at room temperature before cooling to testing temperatures. Mechanical pumping was continued throughout the duration of each test. When it was desired to perform tests with nitrogen present, a nitrogen atmosphere was established within the specimen chamber by repeated vacuum purging with the gas prior to cooling. All tensile data is reported in engineering units. Failure is defined as the first stress maximum in the stress strain curve whether yield or fracture occurs at this point. The stress at this maximum is plotted as a function of temperature.

CRYOGENIC RELAXATION BEHAVIOR

Some linear polymers with in-chain phenyl groups show a board peak or plateau in the 40-50°K region. Hiltner and Baer[1] observed a broad maximum at about 50°K (lcps) in poly-p-xylylene (PPX) which was less intense in an annealed sample. A shoulder at about the same temperature was observed by Roe and Baer[7] in polycarbonate (PC) and a polyimide, poly(4,4'-oxydiphenylene pyromellitimide) (PI). The relaxation spectra of these three polymers are compared with that of PS in Figure 1. The similarity in temperature and shape to the PS δ peak suggests that in-chain as well as pendant phenyl groups may be mobile at cryogenic temperatures.

Cryogenic relaxations in other linear polymers have recently been observed in this laboratory (δ_c peaks). Armeniades and Baer[8] reported two cryogenic loss peaks in polyethylene terephthalate (PET), one at 26°K (lcps) which was dependent on the level of crystallinity, and another at 46°K which was observed in oriented specimens (Figure 2). Similar relaxations in both linear polyethylene (PE) and polyoxymethylene (POM) have been reported by Papir and Baer[9,10]. The complete dependence of these processes on the deformation and thermal history precludes a mechanism involving either specific group motions in the amorphous regions or motions at fold surfaces of the crystal. Rather, that data suggest that the δ_c relaxation arises from the crystalline or ordered regions of the polymer; and a mechanism involving limited segmental motion at dislocations within the crystal has been proposed by Hiltner and Baer[11].

The principal experimental features common to the δ_c relaxation of all three polymers (PET, PE, POM) are summarized as follows:

    (1) The peak is weak or absent in specimens quenched from the melt.

(2) The intensity increases with deformation and annealing.
(3) The temperature of the peak maximum is independent of the pretreatment.
(4) The experimental peak width is slightly larger than that predicted for a single relaxation time.
(5) The activation energy is estimated to be less than 4 kcal/mole.
(6) The relaxation shows pronounced directional anisotropy.
(7) In addition to the primary peak at 45-50°K, PE and POM show a secondary peak at a lower temperature.

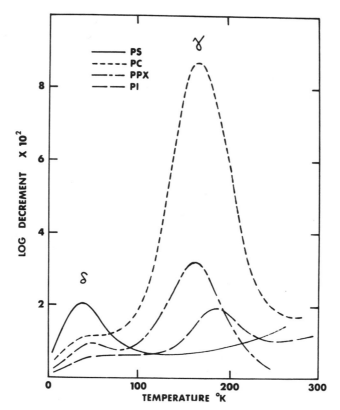

Figure 1.   Relaxation behavior of some polymers with in-chain phenyl groups (lcps). Data from References 1,7.

—————————— polystyrene
- - - - - - - - - - polycarbonate
—·——·—— poly-p-xylylene
— —— — — polyimide

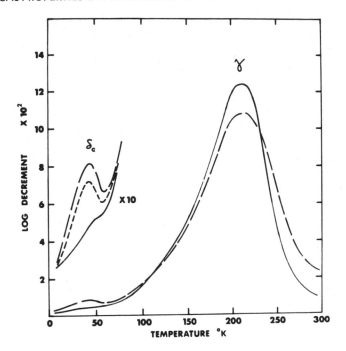

Figure 2.   Effect of orientation on the relaxation behavior of
PET (1cps).   Data from Reference 8.
——————————— amorphous, unoriented
– – – – – – – – – drawn 4X at 40°C
— —— —— —— drawn 5X at 40°C

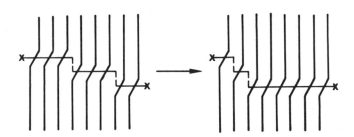

Figure 3.   Redistribution of kinks along a dislocation.

Dislocation motions are known to give rise to relaxation processes in metals and many characteristics of the δ_c peak in PET, PE and POM are similar to experimental features of the Bordoni peaks observed at cryogenic temperatures in fcc metals. The mechanism of the δ_c relaxation proposed by Hiltner and Baer is similar to that proposed by Brailsford[12] for the Bordoni peaks. The model of the δ_c mechanism is based on the assumption that dislocations are present in crystalline polymers as a result of lattice deformation in the vicinity of local stresses. Nonuniform stresses may produce a dislocation with a Burgers vector having a component in the chain direction. In this case, it is proposed that discontinuities or kinks are formed where sections of the dislocation are displaced by one or more lattice spacings in the chain direction (Figure 3). The kinks have an equilibrium distribution along the dislocation which corresponds to a minimum in the free energy. When an external stress is applied, the distribution changes and since the process is thermally activated, gives rise to a relaxation. Movement of the kinks is approximated by a two-state model for the reorientation of the chain segment at the edge of the kink. This abrupt kink model is probably not valid in the case of metals and kinks in metallic dislocations are thought to be better represented by a smooth curve extending over many atomic distances. The abrupt kink model may be better suited for polymeric systems, however, where the number of possible chain conformations is limited by restrictions on bond lengths and angles.

Some of the experimental features of the δ_c peak were outlined above. How these relate to the proposed model will now be considered in more detail.

(1) It has been suggested that the δ_c peak arises from redistribution of kinks along dislocations in ordered or crystalline regions of the polymer. The absence or low intensity of the relaxation in quenched material implies that dislocations are not present in sufficient numbers either because of low amounts of crystallinity (PET) or because the rapid cooling does not permit formation of highly ordered kinked dislocations.

(2) The relaxation is strongly dependent on the deformation and thermal history of the polymer. Since crystalline defects may form in response to any externally applied stress, the intensity of a defect related relaxation process should depend on the entire process history rather than a single parameter such as the amount of deformation or the percent crystallinity. This is observed to be the case. The peak is more intense in oriented samples annealed under constrain than in those that are allowed to relax during annealing (Figure 4). Isotropic, quenched samples exhibit a relaxation subsequent to annealing only if the ends of the sample are held fixed during the heat treatment, no peak is observed if the polymer is unconstrained.

(3) Even in the most highly deformed samples, the intensity of the δ_c peak is almost an order of magnitude lower than the intensity of the PS δ peak which occurs at about the same

temperature. This is consistent with a mechanism in which only a small fraction of the total material, such as the chain segments located at a defect, can participate.

(4) Because the relaxation is thought to be controlled by the motion of a single chain segment, the activation energy, and hence the temperature at which the peak occurs, is an intrinsic property of the dislocation. Therefore parameters which depend on the pretreatment such as the density of kinks in the dislocation or the distance between pinning points, will not be expected to affect the temperature of the relaxation.

(5) The experimental δ_c peak is only slightly wider than that predicted from a single relaxation time. It was seen that the broad spectrum of relaxation times observed for phenyl group motion could be attributed to the large variation in cavity configurations present in the amorphous or disordered regions of the polymer. Crystalline motions, even at the site of a defect, occur within a much more uniform environment and this is reflected experimentally as the relatively narrow spectrum of relaxation times for the δ_c process.

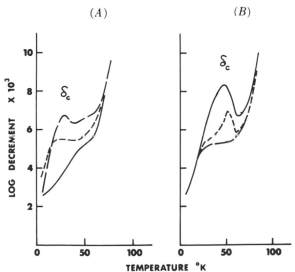

Figure 4.   Effect of pretreatment on the relaxation behavior of
            PET (1cps).   Data from Reference 8.

CRYOGENIC TENSILE PROPERTIES

Stress strain curves for PET in vacuum over the temperature range 4.2-300°K are shown in Figure 5. Above 225°K yielding is observed accompanied by draw band formation. The curves show considerable departure from linearity even down to 45°K, the temperature region of the δ_c relaxation. Below this temperature, the curves are very close to linear.

The plot of failure stress (maximum stress) for specimens tested in vacuum as a function of temperature (Figure 6) more clearly displays the transitions in behavior. This curve can be divided into regimes of behavior on the basis of temperature dependence and failure type (Figure 7). In the case of PET the two commonly encountered regimes of "brittle" and "ductile" behavior are quite apparent. The brittle-ductile transition is indicated at about 160°K by the intersection of the yield stress and fracture stress curves. It must be pointed out that within the "brittle" regime fracture stress is approximately temperature independent. This type of behavior has been observed by Beardmore[13] who in addition observed very little scatter in the fracture stresses of his material, polymethyl methacrylate. He suggested that the fracture process active within this "brittle" regime is an innate property of the material and is not a function

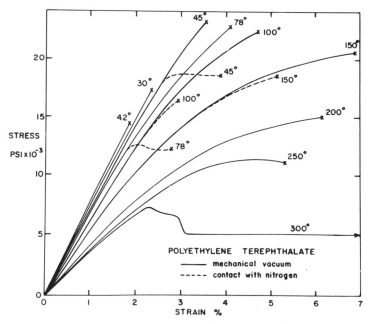

Figure 5   The effect of temperature on the tensile stress strain behavior of polyethylene terephthalate. Shown are results in vacuum and in the presence of nitrogen

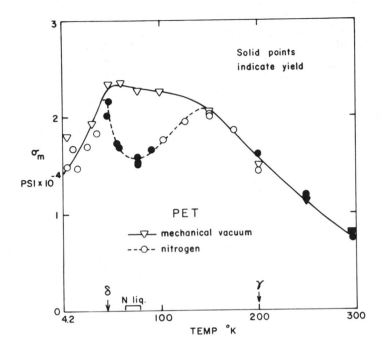

Figure 6   Temperature dependence of the polyethylene terephthalate
failure stress (maximum stress) in vacuum and in the
presence of nitrogen.

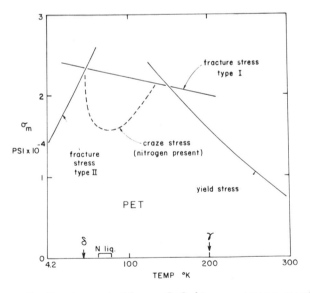

Figure 7   Representation of failure stress regimes.

of variable extraneous stress risers, thus leading to little
scatter and a temperature independent fracture stress. Below
45°K a third regime can be indentified where fracture stress
decreases with decreasing temperature. All possible stress relief
mechanisms must be frozen out at these temperatures, therefore
this third regime is thought to be associated with true brittle
fracture induced by stress risers.

In summary the tensile behavior of PET over the temperature
range 4.2-300°K can be divided into three regimes. Transitions
from one regime to the next can be correlated with the onset of
molecular relaxation modes as revealed by dynamic mechanical
methods. The δ_c process at about 50°K is associated with
departures from linearly in the stress strain curves. The
transition from a temperature dependent fracture stress to a

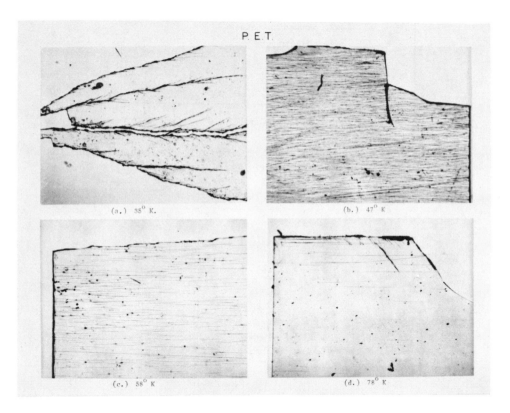

P. E. T.

(a.) 35° K.

(b.) 47° K

(c.) 58° K

(d.) 78° K

Figure 8   Optical micrographs of PET tensile specimens after
           testing at various temperatures with nitrogen present.
                    a) 35°K, bifurcated fracture path, no crazing
                    b) 47°K, numerous small crazes
                    c) 58°K
                    d) 78°K, fewer but larger crazes

temperature independent fracture stress also occurs at the δ_c process. The transition from brittle to ductile behavior is observed in the temperature region of the γ process, 160-200°K, and confirms the results of Roe and Baer[5].

A fourth behavior regime (45-140°K), which is discussed in the following section, can be observed when specimens are tested in the presence of nitrogen. Crazing and apparent yielding characterize the behavior within this region.

Polycarbonate behaves in a manner similar to PET. A brittle region of the type identified by Beardmore is less defined in PC between 30 and 60°K. Below approximately 30°K another region can be distinguished where fracture stress falls off with decreasing temperature. Again as with PET, crazing and yielding is observed between 45 and 100°K for specimens tested in the presence of nitrogen.

AN ENVIRONMENTAL EFFECT

The tensile stress strain curves for PET specimens run in the presence of nitrogen are included in Figure 5. The outstanding phenomena revealed by this data is that nitrogen appears to substantially alter the mechanical behavior of the material particularly in the temperature region 40-100°K. An abrupt yield at low stresses is observed in those specimens tested in the presence of nitrogen. The effect is centered on the liquid state temperature region of bulk nitrogen (63-77°K) and is not exhibited by specimens tested in vacuum. This effect of nitrogen is even more apparent in the plot of failure stress versus temperature (Figure 6). Over a temperature range including but broader than the liquid state region of nitrogen (as marked on the temperature axis of Figure 6) ultimate stresses born by specimens in contact with nitrogen are lower by approximately a factor of fifty percent than stresses achieved in vacuum.

Photographs of specimens tested in the presence of nitrogen are shown in Figure 8. No draw bands are observed in these specimens. Instead they appear heavily crazed. At 45°K, relatively short, thin crazes are formed in high concentration. As the testing temperature is raised to about 80°K, the crazes formed are increasingly longer and thicker though noticeably fewer in number. Above 80°K crazing is comparatively sparce and does not occur at all in specimens tested above 175°K. Specimens tested in vacuum show no crazing below 300°K. At 300°K, PET abundantly crazes prior to draw band formation whether tested in vacuum or nitrogen atmosphere.

The effects of nitrogen on the low temperature behavior of PET and PC are analogous to those of conventional stress crazing agents acting at or immediately above their liquid state regions. The suggestion that nitrogen is a low temperature stress crazing agent has been put forth previously by Parrish[14] and Brown on the basis of their work on PET and PC at 77 and 90°K using lower

strain rates than those employed here. For their materials, Brown
and Parrish observed that crazing and associated yielding in argon
or nitrogen atmospheres could be eliminated by either mechanical
vacuum or helium purge. Natarajan and Reed[15] observed earlier
that natural rubber tested at low strain rates below 150°K in
nitrogen/argon atmospheres yielded concurrent with the formation
of "numerous white striations" believed to be crazes. Interest-
ingly, on subsequent warming to room temperature their specimens
were found to swell into a foam. The gas evolved was mainly
nitrogen and argon, the environmental gases. Finally Billinghurst
and Tabor[16], while studying the effect of hydrostatic pressure on
viscoelastic properties noted that polytetrafluoroethylene was
rapidly plasticized by nitrogen at pressures of a few hundred
atmospheres. Clearly nitrogen (or argon) can not be regarded as
phenomenalogically inert, and perhaps is responsible for many of
the literature reports of low temperature crazing.

It is interesting that the stress crazing effects of nitrogen
are apparent at temperatures both above the boiling point and
below the freezing point of bulk nitrogen. This together with the
observation that crazing effects are eliminated by a brief (3 min.)
pump-down at room temperature (much less time than needed to
achieve diffusion equilibrium), leads to the conclusion that
nitrogen crazing is a surface effect, perhaps entailing adsorption
or capillary condensation.

REFERENCES

1.   A. Hiltner and E. Baer, Crit. Rev. Macromol. Sci., in Press.
2.   J. A. Sauer, J. Polym. Sci., 32C 69 (1971).
3.   J. M. Crissman, J. A. Sauer, and A. E. Woodward, J. Polym. Sci.,
     2A 5075 (1964).
4.   K. M. Sinnott, J. Appl. Phys., 29 1433 (1958).
5.   J. M. Roe and E. Baer, Intern. J. Polymeric Mater., 1 133
     (1972).
6.   C. D. Armeniades, I. Kuriyama, J. M. Roe, and E. Baer, J.
     Macromol. Sci. Phys., B1 777 (1967).
7.   J. M. Roe and E. Baer, Intern. J. Polymeric Mater., 1 111
     (1972).
8.   C. D. Armeniades and E. Baer, J. Polym. Sci. A-2, 9 1345
     (1971).
9.   Y. S. Papir and E. Baer, J. Appl. Phys., 42 4667 (1971).
10. Y. S. Papir and E. Baer, Mater. Sci. Eng., 8 310 (1971).
11. A. Hiltner and E. Baer, Polym. J.,
12. A. D. Brailsford, Phys. Rev., 122 778 (1961).
13. P. Beardmore, Phil. Mag., 19 387 (1969).
14. M. F. Parrish and N. Brown, Bulletin of APS, 17 297 (1972).
15. R. Natarajan and P. E. Reed, J. Polym. Sci. A-2, 10 585 (1972).
16. P. R. Billinghurst and D. Tabor, Polymer, 12 101 (1971).

INDEX

A

Activation energy   249,265
α-chloronaphthalene   286
Amorphous fraction   1,2,101,316
Amorphous pattern   316
Annealing   1, 44
Arrhenius equation   264, 279
Atactic poly(para-biphenyl
   acrylate)   21
Atactic polypropylene   286

B

Bernoulli equation   241
Bisphenol A polycarbonate
   (Lexan)   146
Bordoni peak   340
Bragg's law   45
Branching, long chain   249
Branching, short chain   250
Brittle-ductile
   transition   336
Bucche equation   262

C

Cellulose acetate   307
Cellulose triacetate   309
Chain folding   101, 285
Chromatography, gel-
   permeation   249
Composite structure   2
Continuity equation   245
Copolymer   27, 282

Copolymerization   236
Craze   289, 346
Creep   308
Crosslink   280, 308
Crystalline polymer   1, 43, 285
Crystallinity   21
Crystallization   28

D

Defect   2
Deformation behavior   285
Delta relaxation (or peak) 340
Die swell   239
Dielectric constant   28, 167
Differential scanning
   calorimeter (DSC) 109, 298
Differential thermal
   analysis (DTA)   81
Diffusion   16
Displacement   217
Ductility   141

E

Elastic liquid   239
Elastic modulus   16, 277
Elasticity   239
Electron microscope   4, 48
Energy balance   243
Energy, activation   249, 265
Epitaxial growth   292